Integrated Environmental Management Handbook

Integrated Environmental Management Handbook

Paul W O'Callaghan

JOHN WILEY & SONS
Chichester • New York • Brisbane • Toronto • Singapore

Baffins Lane, Chichester,
West Sussex PO19 1UD, England

Telephone National 01243 779777
International (+44) 1243 779777

Other Wiley Editorial Offices

John Wiley & Sons, Inc., 605 Third Avenue,
New York, NY 10158-0012, USA

Jacaranda Wiley Ltd, 33 Park Road, Milton,
Queensland 4064, Australia

John Wiley & Sons (Canada) Ltd, 22 Worcester Road,
Rexdale, Ontario M9W 1L1, Canada

John Wiley & Sons (Asia) Pte Ltd, 2 Clementi Loop #02-01,
Jin Xing Distripark, Singapore 0512

Library of Congress Cataloging-in-Publication Data

O'Callaghan, Paul W.
Integrated environmental management handbook / Paul W. O'Callaghan.
p. cm.
Includes bibliographical references and index.
ISBN 0-471-96342-9 (cloth)
1. Industrial management—Environmental aspects. 2. Environmental management.
HD30.255.028 1996
658.4'08—dc20 95-52086
CIP

British Library Cataloguing in Publication Data

A catalogue record for this book is available from the British Library

ISBN 0-471-963342-9

Typeset in 10/12pt Ehrhardt by Mayhew Typesetting, Rhayader, Powys
Printed and bound in Great Britain by Biddles Ltd, Guildford and King's Lynn

This book is printed on acid-free paper responsibly manufactured from sustainable forestation, for which at least two trees are planted for each one used for paper production.

To Menna, Catrin, Iwan, Polo and MacMillan

Contents

PREFACE xiii

1 ENVIRONMENTAL PROBLEMS AND REACTIONS: WHY BOTHER ABOUT THE ENVIRONMENT? 1

1.1 Environmental problems 1

Integrated environmental management 2

Definition of the problem 3

1.2 The global perspective 4

Global warming 4

The carbon cycle 7

Correlations and climate models 7

Effects of global warming 9

Fossil fuel combustion 9

Energy conservation 12

Ozone layer depletion 12

Water pollution 13

Animals and vegetation 14

1.3 Mankind 15

World population 15

Energy and fossil fuel consumption 17

Basic needs 17

Calorie 17

Food and drink 18

1.4 Consumerism 18

Transportation systems 18

Energy and environmental costs of manufacturing systems and products 21

'Throwaway' society 21

Reuse and recycling 22

1.5 Sustainable development 22

Global environmental burden 23

1.6 National responses 23

EC Directives 23

The Fifth European Community Environmental Action Programme 24
UK policy 24
UK legislation and regulations 26
Standards 28
1.7 Considerations for companies 29

2 ENVIRONMENTAL LEGISLATION: WHAT ARE THE REQUIREMENTS OF THE LAW? 33
Acronyms 33
2.1 Environmental law 34
Nuisance and duty of care 34
2.2 EC Directives 34
The First Environmental Action Programme (1973) 35
The Second Environmental Action Programme (1977) 35
The Third Environmental Action Programme (1981) 35
The Fourth Environmental Action Programme (1987) 36
The Fifth Environmental Action Programme (1993) 36
2.3 Chronology 38
2.4 Air pollution 38
2.5 Water pollution 38
2.6 Waste on land 42
2.7 Her Majesty's Inspectorate of Pollution 45
2.8 Environmental Protection Act 1990 45
2.9 Penalties 48
2.10 Application for IPC authorisation 48
2.11 BATNEEC 49
Examples of BATNEEC 53
General BATNEEC 58
APPENDIX 2.1 PRESCRIBED PROCESSES AND SUBSTANCES 64
A2.1 Part A processes 64
The production of fuel and power and associated processes 65
Metal production and processing 66
The chemical industry 69
Waste disposal and recycling 70
Other industries 70
A2.2 Part B processes 71
The production of fuel and power and associated processes 71
Metal production and processing 71
Mineral industries 72
Waste disposal and recycling 73
Other industries 73
Prescribed substances 74

3 ENVIRONMENTAL STANDARDS: GOING GREEN? 77
3.1 The European Union's Eco-management and Audit Scheme (EMAS) 77
3.2 Environmental Management Systems: BS 7750 80
3.3 Ecolabelling 83

4 THE INTEGRATED APPROACH TO ENVIRONMENTAL MANAGEMENT: 12 STEPS TO HEAVEN 87
4.1 Integrated environmental management 87
4.2 Environmental issues 90
General issues 91
Process 91
Substances 92
Services 92
Transportation 93
Energy: inventory and energy accounts 93
Emissions to air 94
Water: inventory and water accounts 95
Discharges to water 96
Materials 96
Waste 97
Other nuisances 98
Nature and ecology 98
Accidents, health and safety 99
Products 99
Packaging 100
Information 101
4.3 Definitions 101
4.4 Essential reference material 106
4.5 The chain: an approach to integrated environmental management 107
Step 1: Education 109
Step 2: Commitment 110
Step 3: Initial review 111
Step 4: Policy 112
Step 5: Organisation 115
Step 6: Compliance 116
Step 7: Scoping study 117
Step 8: Management 118
Step 9: Documentation 119
Step 10: Review 120
Step 11: Options/recommendations 129
Step 12: Audit/report 130

4.6 Validation of the environmental management system 138
Independent validation 138
4.7 The 39 sequential steps of integrated environmental management 140

5 ENERGY, EMISSIONS TO AIR AND ENERGY MANAGEMENT 143
5.1 Fossil Fuels 143
End uses for fossil fuels 144
5.2 Combustion 148
Atomic weights 148
Calorific values 148
Organic waste 150
Combustion calculations 151
5.3 Emissions to air 152
Chronology: energy-related emissions to air 152
Prescribed substances 152
The rise of road transport in the UK 159
Pollution from UK industry 164
Non-energy-related emissions to air 165
5.4 Energy management procedure 168
The 39 steps to optimise energy management investments at a site (6) 172
Energy accounting 173
Pollution accounting 179
Energy throughputs 181
Energy uses and options for improvement 182
Rules for the efficient conservation of energy 183
Questions for the client interview 188
Checklists for energy managers 194
Best investments 209
Energy use improvement project 215

6 WATER, EFFLUENTS AND WATER MANAGEMENT 219
6.1 Water supplies 219
The water cycle 219
6.2 Discharges and effluents to water 221
Water pollution 221
6.3 Water and the law 222
Chronology: water and effluents 222
6.4 Water quality 224
Oxygen levels 224
Pollutant content 224
Other properties 224

Emissions to water 231
6.5 Water treatment 234
Drinking water 234
Waste water 236
6.6 The water industry in the UK 237
6.7 Water prices 238
6.8 Uses for water 239
6.9 Water management procedure 239
Water accounting 241
The 39 steps to optimise water management investments at a site 242
Rules for the efficient conservation of water 243
Questions for the client interview 244
Checklists for water managers 246
6.10 Options for water conservation 248
Leakage 248
Water metering 248
Taps 249
Tap flow restrictors 249
Showers and baths 250
Toilets/lavatories 250
Dishwashers and washing machines 251
Rainwater harvesting 251
Grey water recycling 251
6.11 Improvement project 253

7 MATERIALS AND WASTE MANAGEMENT 255
7.1 Materials, products and pollution 255
The 'throwaway' society 256
7.2 Waste and the law 257
Chronology: waste and land pollution 257
Release into land 257
7.3 Waste categories 261
Inert wastes 261
Semi-inert wastes 262
Putrescible wastes 262
Difficult wastes 263
Special wastes 264
Prohibited wastes 265
7.5 Waste arisings 266
7.6 Materials uses and options for improvement 266
Materials and waste accounting 266
Demand side options for cost-effective waste minimisation 267
Energy from waste 270

The 39 steps to optimise waste management investments at a site 274
Rules for the efficient conservation of materials 276
Questions for the client interview 276
Checklists for waste managers 278
Laws of energy and materials flows 280
7.7 Improvement project **281**

8 INTEGRATED CASE STUDY: A CORPORATE SITE **283**
8.1 Energy and environmental audit, Green Products Ltd. **284**
Executive summary 284
8.2 Main report **285**
39 steps of integrated environmental management 285
39 steps of energy management 293
39 steps of water management 312
39 steps of waste management 318
Other studies 331
Aire and Calder 331
Project Catalyst 334

9 SUMMARY **335**
9.1 Summary **335**
9.2 The future **339**

TECHNICAL ANNEX: SUMMARY OF FUNDAMENTAL CONCEPTS, RELATIONSHIPS AND EQUATIONS **341**

BIBLIOGRAPHY **359**

INDEX **363**

Preface

This book is intended as a practical guide to saving resources, whilst saving money and also protecting the environment. The uses of energy, water and materials are considered as the main contributors to gaseous emissions to the atmosphere, effluent discharges to water and the contamination of land by solid and other wastes.

Global environmental problems include the greenhouse effect leading to global warming and climate changes, ozone layer depletion leading to increased ultra-violet radiation reaching the surface of the earth, and water pollution and eutrophication due to excess nitrate run-off. Local environmental problems include traffic pollution from road vehicles, roads and infrastructure; air pollution from vehicles causing respiratory problems and other health hazards; air pollution from power generation and factories; water pollution from people, industry and agriculture; and land pollution from increasing quantities of domestic and industrial waste.

International reactions to these growing problems include the Toronto and Montreal Protocols, which sought to reduce emissions of carbon dioxide and chlorofluorocarbons (CFCs) respectively. Within the European Union, five Environmental Action Programmes have been established to address pollution issues and numerous European environmental Directives have been issued. In the United Kingdom, Her Majesty's Inspectorate of Pollution (HMIP) and the National Rivers Authority were established and new UK legislation and regulations followed, culminating in the 1990 Environmental Protection Act. The Act introduced the concept of integrated pollution control, from which the concept of integrated environmental management emerged.

The British Standards Institution (BSI) produced a specification for Environmental Management Systems (BS 7750) and the European Union introduced the Eco-management and Audit regulation. As a result, companies now have an urgent need to comply with new environmental legislation and so avoid fines and other penalties; to conserve resources by using less input materials and energy and so to save money; to adopt high environmental standards by obtaining BS 7750 accreditation; and/or to comply with the requirements of the EU Eco-management and Audit Scheme.

Integrated environmental management integrates the requirements of environmental legislation; the EC eco-audit regulation; BS 7750; the need to conserve energy, water and materials; the need to reduce polluting emissions to air, effluents to water and solid wastes to land; and the need to save money in the process, via reduced resource and utility costs, reduced effluent and waste disposal charges, reduced pollution taxes, the avoidance of fines and legal penalties and the avoidance of clean-up costs. It thus embraces energy management, water management, materials management and waste management.

The practices and procedures of energy management, water management and waste minimisation are developed in this book. The aim of integrated environmental management is to develop an optimised decision sequence for retrofit conservation options regarding energy, water and waste, whereby a list of management decisions, in order of cost-effectiveness, is formulated. This list leads to an optimal project plan, in which each sequential decision leads to the best action currently available in economic terms. This approach enables environmental managers to adopt 'no-risk' investment policies, in which no further action is taken until financial savings arising from previous decisions have been realised. The plan effectively reinvests 'savings' so that no net capital outlay is required to reduce commodity costs.

Thus the systematic application of integrated environmental management allows all necessary retrofits and modifications to technical activities to be resourced financially from within the environmental management project itself.

An integrated case study of a corporate site, utilising the sets of 39 sequential steps for integrated environmental management, energy, water and waste management developed throughout this book, demonstrates the efficiency, effectiveness and financial value of the approach adopted.

1

Environmental problems and reactions: why bother about the environment?

This introductory chapter reviews and discusses environmental problems, international and national reactions. It defines integrated environmental management and its scope, explains why it is necessary and develops its aims (e.g. to conserve resources and so save money whilst protecting the environment; to avoid legal action, fines and pollution taxes; and to be seen to be environmentally responsible and so to gain a competitive advantage).

1.1 ENVIRONMENTAL PROBLEMS

On Tuesday 1 October 1991, three items of environmental news were reported on the radio:

(1) *'The Greek government is banning cars from the centre of Athens in a drastic attempt to reduce smog pollution.'* In the hot and muggy weather, a huge cloud of poisonous gases had built up to hang above the city. In the absence of any appreciable wind, it had remained for 24 hours. Those prone to respiratory problems were warned not to leave their homes. Thousands of people were suffering and, by 9 o'clock that morning, over 100 had been hospitalised. In the seriously unbalanced *ecosphere* of Athens, a heat island with a pollution dome had developed.

(2) *'Significant levels of insecticide had been discovered in the tissues of polar bears living in Alaska.'* This was attributed to the absorption of these toxic chemicals by sea plankton, eaten by fish, these in turn eaten by seals and the seals eaten by polar bears, the process building in the food chain and concentrating the poisons billion-fold in the bodies of the polar bears.

(3) *'Over one million dolphins were caught in fishing nets each year.'* An ecologist made an appeal for continued research into the development of fishing nets incorporating reflectors to deter dolphins.

Integrated environmental management

Integrated environmental management (IEM) is concerned with the local pollution of air, water and land. It is also concerned with the global effects of pollution, leading to acid rain, ozone depletion, global warming and the eutrophication of water. It is concerned with sea plankton, because it is these organisms which fix vast quantities of the greenhouse gas, carbon dioxide. The sea creatures live, die and sink to the ocean floors, where they remain as carbonates and the organic matter which once formed our fossil oil reserves. IEM is not concerned with the unfortunate deaths of dolphins in fishing nets, although the deaths and diseases of whales, dolphins, seals, fish, shellfish and the like point to poisoning from environmental pollution further up the marine food chain.

IEM is needed to conserve resources, increase utilisation efficiencies, and reduce waste and its associated environmental pollution. Its application leads to effective **integrated pollution control**. It integrates **environmental management** with **environmental engineering** (Figure 1.1), and so covers problems, legislation, standards, economics, policy and impact with respect to energy and air pollution, water and effluents, the use of materials, the production of waste, the production of

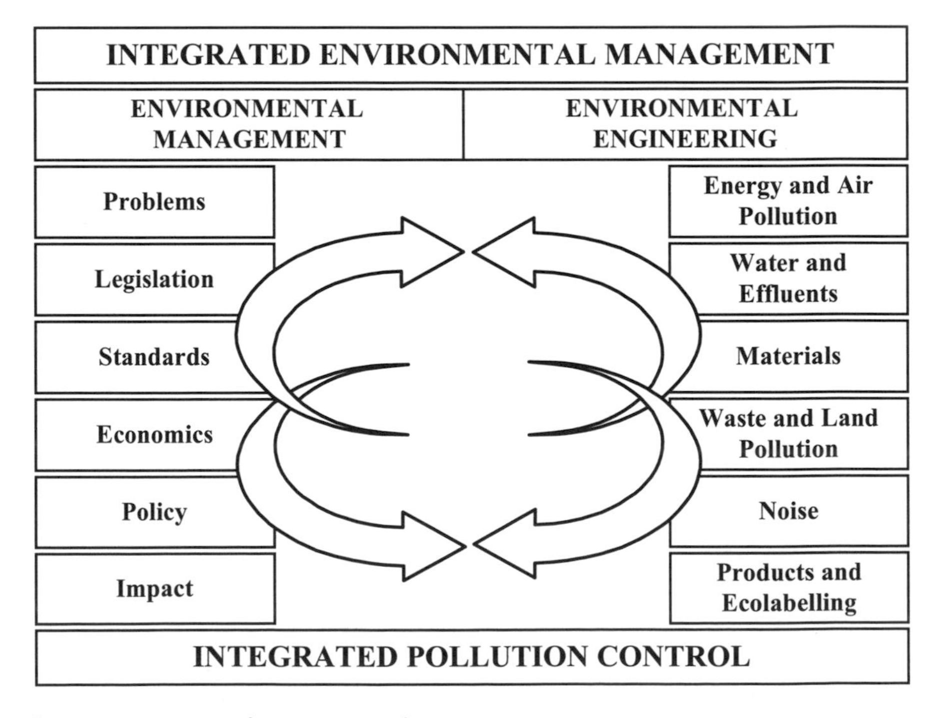

Figure 1.1 Integrated environmental management

land pollution, noise abatement, products and ecolabelling. The chain of activities concerned is developed in Chapter 4.

Definition of the problem

In an attempt to identify the key environmental issues deserving attention, an audience of around 50 people were required to list what they each perceived to be the 10 most pressing environmental problems facing the Earth. A computer analysis of the responses produced the following master list, in order of priority:

Top 10 environmental problems

(1) **Effects of air pollution** (CFCs, heavy gases, car and truck pollution, pollution from factories, animal flatulence, greenhouse effect, global warming, acid rain, ozone layer depletion).
(2) **Water pollution** (nitrates, water quality, oil pollution at sea, effects on marine life).
(3) **Agricultural effects** (deforestation, rainforest destruction, drought, desertification, overfarming, nitrates, leaching, change of land use, flooding).
(4) **Population growth**
(5) **Land pollution** (growth of domestic and industrial waste, disposal problems, dumping).
(6) **Mankind** (world economic policies, education, poor attitudes, ignorance, short-sighted attitudes, greed, selfishness, social pressures, religion).
(7) **Nuclear (chemical) waste and war**
(8) **Depletion of resources** (exploitation of fossil fuels against clean energy, built-in obsolescence).
(9) **Low living standards** (lack of sanitation, health, famine, starvation).
(10) **Ecological effects** ('gene pool' destruction, lack of biological diversity).

It appeared that these concerns could be classified into **global environmental problems** and **local environmental problems**, as follows.

Global environmental problems

- The greenhouse effect leading to global warming and climate changes
- Ozone layer depletion leading to increased ultra-violet radiation reaching the surface of the Earth
- Water pollution
- Population growth leading to famine

- Nuclear explosions
- Depletion of the world's resources
- Destruction of biological diversity

Local environmental problems

- Traffic pollution from road vehicles, roads and infrastructure
- Air pollution from vehicles causing respiratory problems and other health hazards:
  - unburnt hydrocarbons
  - carbon monoxide from incomplete combustion
  - oxides of nitrogen (NOxs) which form nitric acids and so contribute to acid rain (acid rain kills fish and damages trees, other vegetation, habitats and buildings)
- Ozone (so called 'low level' or 'ground level') from the action of sunlight on exhausts
- Air pollution from power generation and factories
  - soot
  - smoke
  - unburnt hydrocarbons
  - carbon monoxide from incomplete combustion
  - oxides of nitrogen (NOxs) which form nitric acids and so contribute to acid rain
  - sulphur dioxide, which forms sulphuric acid and so contributes to acid rain
- Water pollution from people, industry and agriculture
- Land pollution from increasing quantities of domestic and industrial waste (and leachate to watercourses)
- Disposal of nuclear and other toxic waste

1.2 THE GLOBAL PERSPECTIVE

Without doubt, the most concerted international responses to environmental problems have been stimulated by fears of global warming, global ozone layer depletion and water pollution.

Global warming (1)

Long-wave radiation from the Earth escapes through an infra-red window which occurs in the near infra-red at 6.5 to 14 μm. It is said that the greenhouse

Table 1.1 The greenhouse gases

Carbon dioxide	CO_2	from the combustion of fuels, respiration and decay
Methane	CH_4	from animal flatulence, landfill gas, natural gas leaks, rice paddies and swamps
Chlorofluorocarbons	CFCs	from aerosol sprays, refrigerants, packaging foams and solvents
Nitrogen oxides	NOxs	from fuel combustion and fertiliser use
Ozone	O_3	from reactions involving pollutants (e.g. photochemical smog)

Table 1.2 Breakdown of the sources of the greenhouse gases (CO_2 equiv. by mass, ppmm, over pre-industrial times)

	1985 Total	CFC effect	Food production	Fossil fuel	Business as usual 2050
CFCs	40 (19%)	40			2290 (57%)
CO_2	122 (57%)			122	505 (13%)
CH_4	27 (13%)		27		69 (2%)
N_2O	6 (3%)		3	3	21 (0.5%)
O_3	20 (9%)	20			1149 (28%)
Total	215 (100%)	60	30	125	4034 (100%)

gases, mainly carbon dioxide (CO_2) at present, are filling this hole through the troposphere.

The 'greenhouse effect' is a thermal radiation rectifier whereby incoming short-wave solar radiation is allowed to pass through this blanket of 'greenhouse gases' in the Earth's troposphere, whilst the passage of long-wave radiation from the Earth's surface to the vault of deep space is inhibited by the absorption characteristics of these gases.

In the absence of the greenhouse effect, the mean temperature of the Earth would be some 33°C colder than it is at present (15°C).

The greenhouse gases are listed in Table 1.1. Human activity releases pollutants into the troposphere, and these give rise to the 'greenhouse gases', which are threatening to disturb the thermal balance of the Earth to produce a new equilibrium state. In 1985, the aggregated breakdown of the sources of greenhouse gases was as given in Table 1.2 (2).

It has been estimated (3) that carbon dioxide might contribute about 50% to global warming. Methane, nitrous oxide, tropospheric ozone, water vapour and the CFCs may thus currently double the effect of CO_2, as Table 1.2 indicates. These figures for 1985 are broadly in line with those estimated by other workers (4,5).

The build-up of greenhouse gases is cumulative and it is said that, since

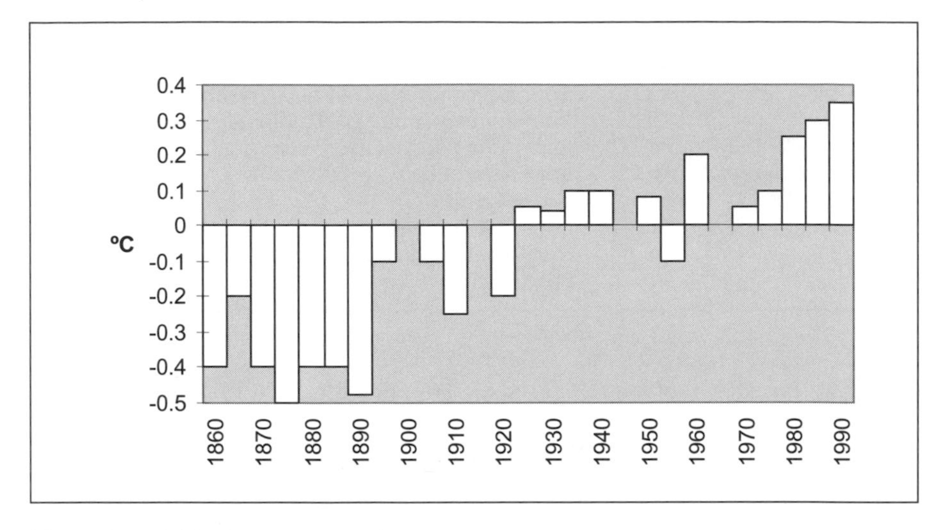

Figure 1.2 Global temperature rise

pre-industrial times, concentrations have increased significantly. For example, measurements taken between 1973 and 1985 at Alaska, Hawaii, Samoa and the South Pole (reported in reference (6)) indicate that atmospheric carbon dioxide concentrations rose on average from 327 ppmv (parts per million by volume) to 345 ppmv over that period (1.5 ppmv/annum). Gribbon (6) extended this data to show that the trend continued to 350 ppmv in 1989. Analyses of air bubbles trapped in glacial ice (6,7,8) have indicated that carbon dioxide concentrations were stable at 270–280 ppmv over the period from 10 000 years ago to the nineteenth century.

It was concluded, therefore, that the proportion of carbon dioxide in the Earth's atmosphere increased from 280 ppmv pre-1800 to 350 ppmv today, and that it continues to increase at 1.5 ppmv each year.

Concentrations of other greenhouse gases are also building up in the atmosphere: methane, nitrous oxide, tropospheric ozone and halocarbons (CFCs). Any change in the composition of the tropospheric atmosphere changes the thermal balance of the Earth.

Researchers at the University of East Anglia (6) have compiled a record of measured global temperatures from 1850 to the present day from weather reports, rainfall and temperature records and indications of past climatic conditions locked in trees, ice cores and rocks.

Whilst the data (Figure 1.2) suffer from seasonal cyclic variations, there is little random scatter and the increasing trend of temperature is clear. Although natural variations in climate occur from changes in the sun's luminosity, volcanic eruptions, random factors and long timescale systematic drifts, the data suggests that an overall mean global temperature rise of 0.5°C has occurred since 1850 (pre-industrial times).

The carbon cycle

The *carbon cycle* is the circulatory chain of events by which carbon is exchanged between animals and plants and the global environment. Atmospheric carbon dioxide is absorbed by plants via the process of *photosynthesis*, by which carbohydrates are formed in living plants by the action of sunlight on chlorophyll. This absorption occurs directly from the air through the underside of the leaves, or indirectly from water via the roots. Oxygen is formed and released to the atmosphere in the process.

Respiration by animals is the process by which oxygen is transported from the lungs to the muscles by the haemoglobin in the blood. The carbohydrates oxidise to form carbon dioxide which is transported back to the lungs to be expelled to the atmosphere. Respiration is thus the opposite process to photosynthesis, carbohydrates being converted to carbon dioxide, water and chemical energy by living creatures.

Plants are the only living things able to manufacture their own food internally, using photosynthesis to convert water, carbon dioxide and minerals into sugars, proteins and other energy sources. During photosynthesis, energy from the sun is converted into food energy in the form of glucose. To liberate this energy, this glucose reacts with oxygen to form carbon dioxide and water. Thus plants also respire continuously. The chemical reaction involved is described by the respiration equation (see the Technical Annex), so oxygen is absorbed.

The carbon 'fixed' by photosynthesis in plants and animals is eventually returned to the atmosphere as carbon dioxide when plants and animals die and the dead organic matter is consumed by the decomposer mechanisms.

Atmospheric carbon dioxide is washed down by rain to form carbonic acid in sea water. Phytoplankton are single-celled organisms that take up carbon dioxide from sea water and, by photosynthesis, use it to produce carbohydrates and oxygen, which dissolves in the water. Zooplankton and fish consume the carbon fixed by the phytoplankton. When these predators and uneaten phytoplankton die, the carbon is returned as carbon dioxide dissolved in seawater. Shellfish fix carbon as carbonates in their shells, which are eventually discarded to the ocean floor to form chalk, limestone and marble sediments. If these substances are heated, carbon dioxide is released to the atmosphere, completing the oceanographic carbon cycle.

Correlations and climate models

The argument for 'global warming' says that the extra CO_2 released from fossil fuel combustion, over and above that exchanged between animals and vegetation in the natural carbon cycle, is enough to disturb the thermal equilibrium of the Earth, resulting in higher mean temperatures. By projecting forward rates of greenhouse gas emissions, and correlating these with past temperature rises, computer models predict that the global mean temperature will rise by between 1.5°C and 4.5°C by the year 2050.

These predictions represent the highest temperatures for 100 000 years. The last ice age (18 000 years ago) was associated with temperatures only 4°C colder than those prevailing today. The change in global temperature at the end of that ice age was accompanied by a doubling of CO_2 levels in the atmosphere, thought to be caused by a decrease of plankton in the oceans.

Climate models are based upon finite difference solutions of global meteorology. Atmospheric general circulation models are based upon the equations which determine the motion of the atmosphere (wind), its thermodynamics (temperature) and conservation of water (humidity and surface pressure). The values of these basic variables are kept at various locations over the globe and at various levels in the atmosphere. A nodal point is typically 5 degrees of latitude by 7 degrees of longitude.

The major uncertainties in the predictive models are the effects of the ozone hole in the stratosphere, the speed of warming, the response of oceans and ice and the vegetable and animal responses. Global mean temperature change is due to the combined impacts of (i) the greenhouse effect of carbon dioxide and the other trace gases; (ii) the increased reception of short-wave radiation from the sun, resulting from the destruction of the ozone layer in the stratosphere; and (iii) the reflection of sunlight by sulphuric acid droplets thrown up by volcanic eruptions and fossil fuel combustion. The individual effects are difficult to quantify owing to the sparse and often conflicting data.

If the 0.5°C global temperature rise were due solely to an additional 80 ppmv of carbon dioxide directly added to the atmosphere, the global temperature and concomitant sea level rise could be correlated to a first order with cumulative fossil fuel consumption.

Thus a simple analytical predictive which correlates temperature rise with cumulative fossil fuel consumption, neglecting the other greenhouse gases (CFCs, methane, nitrous oxide, subtropospheric ozone and water vapour), yields predictions similar to those indicated by the complex models, because these treat these other gases in terms of their *global warming potential* with respect to carbon dioxide.

Data has been extrapolated (2) to indicate that, if world fossil fuel consumption continued to increase at 2.79%/annum (i.e. as in 1987), reserves would be depleted by 2050, the global temperature would rise to 17°C and sea levels by over 0.4 metres.

This method of projection also assumes that the proportion of carbon dioxide fixed by vegetation remains constant at its present level (24%) over the period of the projection. At some unknown level of fossil fuel carbon dioxide emissions, however, the process of photosynthesis might become saturated, especially as large-scale deforestation continues. At the time of saturation, atmospheric carbon dioxide concentrations, global temperatures and sea levels would begin to rise at four times the predicted rates, resulting in up to 8°C global temperature rise. Secondary effects were not included, such as the reduced absorption of carbon dioxide with increasing temperature by sea water (at 20°C, the solubility of CO_2 in water reduces by 10% for every 1°C temperature rise) or the reduced reflectivity of the Earth as ice melts, exacerbating the problem, or the reduced energy needs associated with increased environmental temperatures (e.g. reduced space heating requirements), helping to reduce the problem.

The situation could be much worse if the rates of production of the other greenhouse gases increase, or even remain at the current levels. CFCs are not fixed by vegetation. McElroy (3) predicts that the greenhouse gases other than carbon dioxide, and particularly the CFCs, will contribute an additional 180% to global warming in the 2020s (see also Table 1.1).

Consensus

> Most scientists agree that global warming is inevitable and that we can only delay temperature rises to buy time. (6)

Effects of global warming

The effects of global warming have been projected to indicate sea level rises, more extreme local temperatures (especially in the northern latitudes), climate changes, changes in precipitation rates, increased rates of desertification, the more frequent occurrence of extreme weather events (droughts, floods, storm surges and hurricanes), adverse agricultural effects, increases in algae growths, increases in the number of pests and tropical diseases, and food and water shortages.

It has been suggested that, as the world has become warmer, the water in the seas has expanded and there has been some glacial melting (7), which has resulted in a rise in mean sea levels of 0.1m over the period from 1850 to 1987.

There are, however, many counter arguments to these theories. Weighing the arguments for and against the occurrence of global warming results in significant uncertainty and no firm conclusion as to whether human activities are leading to significant global climate change can be deduced from the data available (2).

Nevertheless, the *precautionary principle* suggests that a global environmental management programme should be instituted.

Fossil fuel combustion

Burning organic materials (fossil fuels, wood, plants and animal matter) speeds up the process of decay, releasing energy and carbon dioxide in the process (cf. Appendix 2.1).

So, carbon dioxide enters the atmosphere via respiration, fermentation, decay, the burning of vegetable matter and the combustion of fossil fuels. It is 'fixed' to carbon via photosynthesis in land and sea plants and is dissolved in the seas. Until industrialisation, the amount of carbon being fixed each year exactly balanced the amount being released, hence the constant 270 ppmv prior to industrialisation (6). Since then, carbon dioxide which was fixed in the fossil fuels over a period of 28 million years has been released into the atmosphere via the combustion of fossil fuels. With the growth in the consumption of fossil fuels, some of the excess carbon dioxide has not been fixed or dissolved but has remained in the atmosphere (6).

Carbon dioxide emissions are said to be causing global warming. If so, global carbon dioxide emissions should be curbed by reducing rates of fossil fuel combustion and by major reafforestation programmes. Carbon dioxide is produced in the combustion of all combustible materials and is expelled by all animals and vegetation in the processes of respiration and decay.

There is no doubt that sulphur dioxide and NOx gases cause respiratory problems and form acid rain, which kills living matter and damages buildings. Sulphuric acid droplets in the upper atmosphere reflect away solar radiation and thus have a cooling effect on the Earth. Volcanoes throw up vast amounts of sulphur dioxide. Some protagonists of the global warming theory suggest that the sulphuric acid droplets created by the eruption of mount Pinatubo in the Philippines counteracted the incessant rise in global temperatures for a time. Past temperature records are indeed greatly effected by volcanic eruptions. Sulphur dioxide is produced by burning crude oil (0.35–4.15% sulphur), petrol (0–0.08% sulphur), fuel oil (0.35–4.93% sulphur), coal (0.3-3.8% sulphur), lignite (0.9% sulphur) or orimulsion, the new bitumen-based and invariably filthy fuel, which contains 4.85% sulphur (9).

Combating global warming

In 1988, the **Toronto Protocol** called for a reduction of carbon dioxide emissions to 80% of 1988 levels by 2005 as a global goal to stabilise atmospheric levels.

The European Commission has indicated a desire to stabilise carbon dioxide emissions at their 1990 level by the year 2000. In order to achieve this, energy saving measures are being encouraged and **carbon taxes** have been proposed (see the Technical Annex), at £18/tonne of carbon emitted in 1993 to £60/tonne of carbon emitted in the year 2000.

Ensuing price rises by the year 2000 for the energy commodities would be as shown in Table 1.3.

It is not clear whether power generators will be liable to pay carbon taxes for carbon dioxide releases during electricity generation, but these proposed taxes are opposed by power generators, industrialists and many politicians on the basis that Japan and the United States should adopt similar courses of action so that prevailing international market factors are not distorted.

Table 1.3 Price rises for the energy commodities as a result of applying carbon taxes

Commodity	p/kWh	p/unit
Natural gas	0.324	0.5 p/therm
Coal	0.768	£50/tonne
Oil	0.414	5.8 p/litre
Electricity	2.304	2.3 p/kWh

Nevertheless, the British government has stated that it is prepared to set itself the target of returning carbon dioxide emissions to 1990 levels by 2005. It has introduced value-added tax (VAT) on domestic fuel as a means to assist this.

Note that the imposition of carbon taxes, rather than sulphur taxes, emanates from a belief that global warming, stemming from greenhouse gas build-up from fossil fuel combustion and leading to climate change, is a serious and imminent threat.

More immediately dangerous to human health are the low-level local pollutants: unburnt hydrocarbons, carbon monoxide, nitrogen oxides, low-level ozone, soot and smoke, and sulphur dioxide, which cause respiratory and other health problems.

Diesel engines using high compression ratios are undoubtedly more efficient in converting fuel to motive power, reducing the amount of carbon dioxide produced per kilometre travelled, but they produce more black smoke and particulates, adding to local pollution.

Catalytic converters for cars convert unburnt hydrocarbons and carbon monoxide into carbon dioxide and water vapour, and also convert nitrogen oxides into nitrogen and water. The net effect is a reduction in local toxic emissions which lead to smog and acid rain, but there is a concomitant increase in carbon dioxide release, which is amplified by the need to burn more fuel per kilometre to overcome the additional pressure drop imposed by the converter.

Flue gas desulphurisation equipment, installed at power station boilers, absorbs sulphur dioxide from the flue gases into a limestone slurry spray to form gypsum (calcium sulphate). Its introduction requires a greater specific fuel consumption to overcome the extra energy requirements of the equipment and so leads to the production of greater quantities of carbon dioxide.

It is therefore possible to envisage expensive combinations of *environmental performance improvement measures*, which cancel one another out depending upon the relative significance and weighting factors assigned to the various types of emission. There are even some global warming protagonists who believe that sulphur dioxide *should* be produced and emitted to counteract the global warming due to greenhouse gas emissions!

Global and local problems should, however, be prioritised, so that correct courses of action can be taken. For example, *should there be sulphur and NOx taxes*?

Energy conservation

Leaving aside the greenhouse effect and the possibility of global warming, the needs for ecologically safe and clean local environments, pure air, water, land, food and life forms, and for the conservation of energy and mineral resources, whilst avoiding pollution, poisons, unnatural products and unnecessary waste, make energy and environmental conservation desirable. As most environmental pollution stems, directly or indirectly, from fossil fuel combustion, energy conservation is seen as a priority area for attention. Furthermore, **energy savings result in financial savings and also in reduced environmental pollution**.

It will be demonstrated later that water and materials conservation and waste reduction also save money by reduced input resource costs, reduced charges for effluent and waste disposal, reduced pollution taxes and the avoidance of fines and other penalties for pollution offences. Competitiveness is also enhanced by a company being perceived to be clean and green.

If local pollution problems are given attention, then global problems will be solved:

Think global
Act local

Carbon dioxide emissions may be reduced via flue gas clean-up and energy conservation. Flue gas clean-up is an expensive option. Energy conservation brings with it the added benefit of financial savings.

Ozone layer depletion

The Earth's 'ozone layer' is a layer of ozone (O_3) in the stratosphere formed by the action of sunlight on oxygen. Solar radiation incident at the perimeter of the Earth's surface contains 5% ultra-violet radiation. The ozone layer helps to reflect most of this short-wave radiation so that the modified radiation spectrum reaching the surface of the Earth contains only 1% ultra-violet radiation.

Chlorofluorocarbons (CFCs) reaching the stratosphere, under the action of sunlight, convert ozone to chlorine (Cl) and oxygen. This chlorine breaks down further ozone molecules so that one chlorine atom destroys 100 000 ozone molecules before it disappears in atmospheric chemistry (1). Halons (fire-fighting gases) contain bromine, which has a similar effect and could be up to 50 times more damaging to the ozone layer than chlorine. Other ozone-depleting chemicals include the cleaning agents methyl chloroform, carbon tetrachloride and trichloroethane.

The situation is exacerbated by the chemicals thrown up by volcanic eruptions. These destroy nitrous oxides, which normally help prevent ozone destruction in the atmosphere. Hence the increasing loss of the Earth's protective ozone layer, which could be as great as 30% by the year 2000, resulting in increased amounts of ultra-violet radiation reaching the surface of the Earth. The amount of ultra-violet light reaching the ground in Europe has been increasing at about 1% per annum since 1980.

Ozone layer depletion also has a cooling effect on the Earth's atmosphere, allowing greater rates of heat loss through the stratosphere and hence somewhat counteracting global warming due to the presence of CFCs in the troposphere.

The effects of exposure to ultra-violet radiation are cumulative, causing skin cancer, eye cataracts and weakening the body's immune system. Remedies include wearing sunglasses, avoiding skin exposure and using sun creams. Some scientists also agree that crops are damaged, agricultural yields are reduced, and sea plankton (CO_2 absorbers and the base of the aquatic food chain) are killed off by greater ultra-violet radiation levels.

Combating ozone depletion

In 1987, the **Montreal Protocol** called for a reduction in the world production of CFCs to the 1986 level in 1989, 80% of this level in 1993 and 50% of this level in 1998. The European Union is committed to phasing out all ozone-depleting chemicals by the year 2000, but all member countries wish to bring this target date forward to 1995. Britain is committed to phasing out CFCs by 2000 but wishes to bring this date forward to 1997.

Water pollution

The water cycle

Regular supplies of rain are essential to the operation of all the cycles of life. In *the water cycle*, the evaporation of water from rivers, lakes and seas provides water vapour which is held in the atmosphere as clouds. When the clouds cool, the water vapour condenses and falls as rain.

The nitrogen cycle

All living organisms need nitrogen to build up proteins, essential ingredients for all living matter. The *nitrogen cycle* is the process by which atmospheric nitrogen enters living organisms, being absorbed into green plants in the form of nitrates, the plants being eaten by animals, and the nitrogen being returned to the ecosystem through the animal's excreta or when a plant or animal dies (10).

The amount of biological fixation of nitrogen on Earth has been estimated as being of the order of 40 million tonnes per annum (11). Simultaneously, 40 million tonnes are returned to the atmosphere each year by denitrifying bacteria, bringing the global system into balance.

This natural cycle has now been disturbed by the need to produce ever-increasing supplies of food. If crops are constantly harvested from the same plot of land, and vegetation is not left to decay on the spot, the soil becomes low in nitrogen compounds. Over the past 40 years, the amount of nitrogen fertiliser needed to produce the same yield per acre has increased more than five fold.

Nitrogen is fixed artificially by the Haber Process, which operates at 500°C and 200 atmospheres. Methane (energy) is reformed to produce hydrogen, which reacts with nitrogen and oxygen to form carbon monoxide. A catalytic reaction combines nitrogen and hydrogen to form ammonia. This is converted to nitric acid and then the ammonium nitrate fertiliser, which is fed to the soil to supplement the natural cycle. Current global industrial nitrogen fixation rates are around 40 million tonnes per annum—of the order of those occurring in the natural cycle (11). In addition to this, world artificial fertiliser production currently doubles every 10 years.

Since they cannot be recycled to the atmosphere by the denitrifying bacteria, the manmade nitrates eventually end up as run-off (via fertilisers, sewage and other waste) into lakes, rivers and seas, where *eutrophication* (accelerated ageing by stimulating the growth of oxygen-hungry algae blooms) can result.

In the seas these algae destroy phytoplankton, which are important carbon dioxide absorbers and oxygen producers. Thus pollution of water by manmade nitrates **accelerates the greenhouse effect**.

The phosphorus cycle

Phosphorus is vital for both plants and animals. It is required by plants to produce good root systems and it is present in bones and nerve tissue.

Artificial phosphates used in agriculture are highly concentrated. Phosphates escape into watercourses from sewage, especially if the waste water contains detergents, and encourage the growth of algae, **promoting eutrophication**.

Animals and vegetation

For life to exist on Earth, the cycles of life must continue to operate.

Vegetation takes in oxygen and gives out carbon dioxide in the process of photosynthesis, whilst animals take in oxygen and give out carbon dioxide in the process of respiration. The carbohydrates in the body emanate from food, which originates directly or indirectly from plants. Atmospheric nitrogen enters living organisms, being absorbed into green plants in the form of nitrates, the plants being eaten by animals to build proteins. Nitrogen is returned to the ecosystem through the animal's excreta or via decay. Phosphorus is needed by plants to produce good root systems. The plants are eaten by animals, producing bones and nerve tissue. When animals die and decay, phosphorus is returned to the soil. Cycles for calcium, iron, potassium, manganese, sodium and sulphur involve similar exchanges between plants and animals.

Animals cannot exist without vegetation, although vegetation could exist without animals.

1.3 MANKIND

World population

Table 1.4 (10) and Figure 1.3a show estimates of world human population since 8000 BC. It may be seen that, by the year 2000, the number of human beings on Earth will reach 6000 million, the number currently doubling every 35 years. Table 1.4 also shows that **the total number of people alive today exceeds the total number of human beings that have lived and died on this planet in the past.**

According to the University of Rhode Island World Hunger Project (12), with present levels of agrotechnology and if all human beings were strict vegetarians, the

Table 1.4 Estimated human population growth since 8000 BC

Year	Millions of people	Time (in day span equivalent)	Event
8000 BC	6	0.19 seconds ago	Advent of agriculture
1 AD	255		
1000	255		
1250	400		
1500	460		
1600	600		
1700	680	4.35 milliseconds ago	Industrial Revolution
1800	950		
1900	1600		
1920	1860		
1930	2070		
1940	2295		
1950	2500		
1960	3050		
1970	3700		
1980	4400		
1985	4800		
1987	5000		
projected			
2000	6000		
2025	8200		
2090	10500		

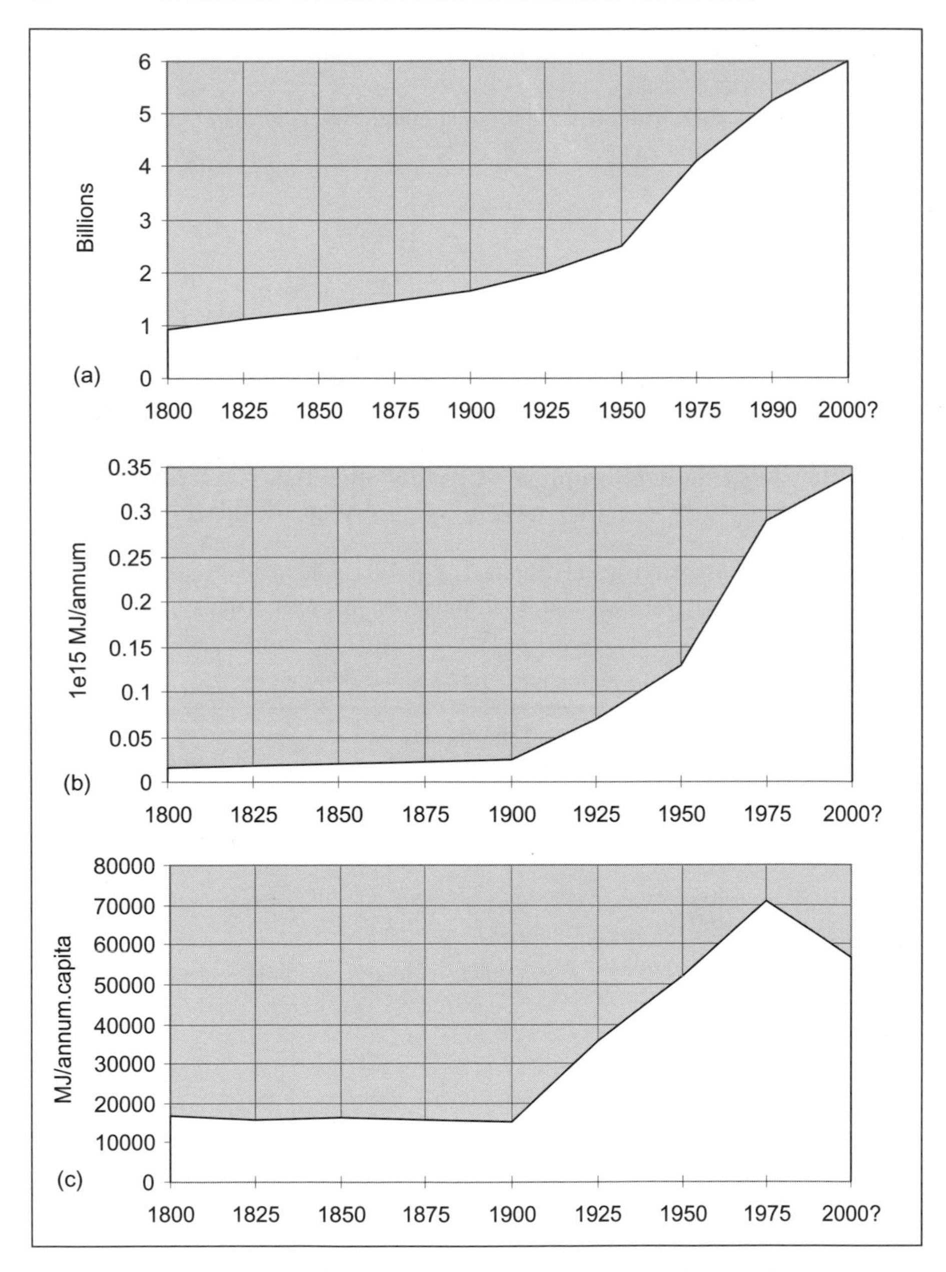

Figure 1.3 (a) World population; (b) World energy consumption; (c) Energy consumption per head of world population

Earth could support a maximum of 6000 million people. If 15% (e.g. as in South America) or 25% (e.g. as in North America and parts of Europe) of the world human diet were, in calorie terms, derived from animal protein, this maximum drops to 4000 or 3000 million respectively. Without the use of fossil-fuel-derived energy in agriculture mainly for the production of artificial fertilisers, the Earth could support considerably less people (13).

Energy and fossil fuel consumption

Figure 1.3b shows the rise in world energy consumption. Prior to the Industrial Revolution, the only fuel used was wood for cooking and heating. The significant use of the fossil fuels does not show on the graph until around 1875, thereafter rising exponentially until the rate of growth of energy use begins to fall back from 1975 onwards.

Figure 1.3c plots the energy consumption per head of world population. This rises constantly from 1900, but experiences a peak in 1975 and then starts to drop off. For the first time in the history of the world, energy consumption is now rising more slowly than population growth.

Basic needs

The basic needs of human beings are food and drink, clothing and shelter. In a civilised society, education and health care might also be expected as basic requirements.

Probably the smallest indivisible unit capable of being audited in terms of its environmental impact is a single animal, which takes in air, food and water, metabolises to harness (originally solar) energy in its muscles, expends work, 'burns' fat and food, rejects heat and water, expels carbon dioxide and excretes water and solid matter.

Over a period of time, materials and energy inputs and outputs are in balance.

Calorie

The unit used in quantifying the energy content of food is the 'calorie', defined as the amount of energy required to raise the temperature of 1 lb of water by 4°F.

1 lb = 0.454 kg
The specific heat of liquid water at 20°C = 4182 J kg^{-1} K^{-1}
and 1°F = 5/9°C = 0.555°C
and so, 1 Calorie = m s ΔT = 0.454 $\times$ 4182 $\times$ 0.555 $\times$ 4 = 4215 J

Food and drink

For self-sufficiency, Michael Allaby (14) has calculated that a single person would need 5000 to 10 000m^2 of agricultural land. Although this is currently almost exactly that which exists on a global scale on Earth, some countries have more growing land than others. Hong Kong has approx 260m^2 per capita, whilst French Guinea has some 2 000 000m^2 per capita. A country which has less than 10 000m^2 per capita of agricultural land either starves or manufactures goods for export in exchange for food. It is interesting to note that at the time of the Industrial Revolution, Britain's population was such that there was approximately 10 000m^2 of agricultural land per capita.

1.4 CONSUMERISM

Figure 1.4 illustrates the problem of the consumer society. Raw materials and resources are fed into the upper pipe, which processes these into products. This pipe exudes money, employment for consumers and profits. The consumer pays the money earned to feed from this pipe, derives satisfaction from the experience and discharges its waste to the lower pipe, which ejects pollution to the environment. The process is continuous. The faster the throughput, the more the money flows, the faster the rate of consumption and the greater the level of consumer satisfaction.

Any obstacle or blockage in the two-pipe system, such as resource conservation or pollution prevention, dries up money, employment and profits, as well as consumer satisfaction.

It has been seen that, because of this accelerated consumerism, resource utilisation and polluting emissions to air, water and land, the Earth's protective ozone layer is being destroyed, global warming may be occurring and the rivers, lakes and seas are dying.

Transportation systems

In the UK, USA and Europe, a major proportion of all energy and materials is consumed in road and rail building, bridge construction, maintenance, vehicle manufacture, garages, vehicle repairs, petrol and diesel fuels, and the direct use of electricity for transportation. More roads lead to more suburbs, more suburbs lead to more commuting, More commuting leads to more traffic, more traffic needs more roads . . . !

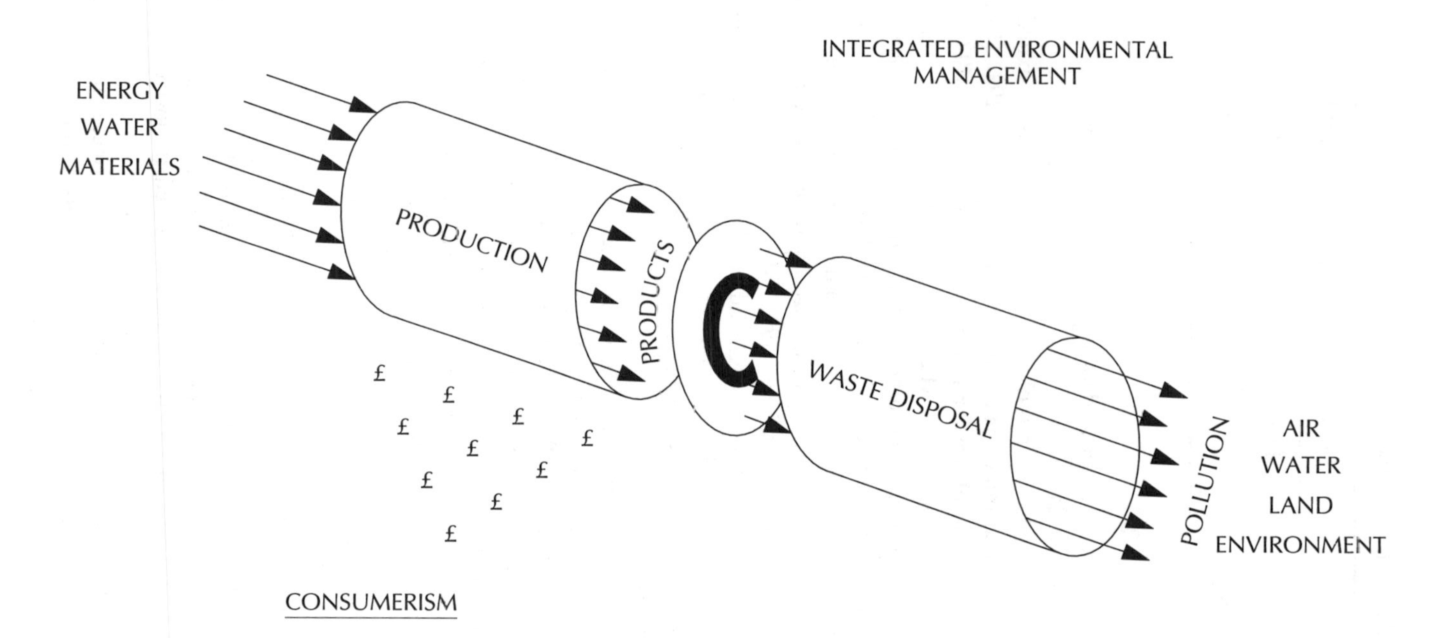

Figure 1.4 The consumer society

Table 1.5 Forecasts of motor vehicles on British roads—vehicle-km (1990 = 100)

Mode	1995	2005	2015	2025
Cars	110	140	165	190
Heavy goods vehicles	108	120	160	190
Light goods vehicles	110	145	185	245

More roads lead to distributed manufacturing sites, these lead to more traffic, which needs more roads . . . !

In the UK the volume of traffic on the roads increased sixfold and car use (vehicle-km) increased more than tenfold between 1950 and 1991 (15). Final users' energy demand for petroleum for transport doubled from 19 MTOE (megatonnes of oil equivalent) in 1960 to 38 MTOE in 1985. In 1960, 25% of transportation energy was provided from coal to power steam trains, with petroleum for transport comprising 16% of the final users' energy demand. In 1985, this had increased to 25%, by 1991 to 31% (16) and it is currently approaching 35%.

As a result, UK concentrations of nitrogen dioxide, mainly from motor vehicles, increased by 35% between 1986 and 1991, NOx emissions increased by over 70% between 1980 and 1993 (17), background levels of ozone have also risen substantially, emissions of volatile organic compounds (VOCs) were 4% higher in 1992 than in 1980, and carbon monoxide emissions (90% due to road vehicles) had increased by over 30% since 1980 (15).

Between 1970 and 1990, total carbon dioxide emissions by end-user fell from 180 million to 160 million tonnes, due mainly to a 28% contraction in industrial and agricultural activity and a 10% reduction in domestic emissions. Carbon dioxide emissions from road transport, however, increased, from 19 million tonnes in 1970 to 34m tonnes in 1990 (+79%), counteracting the gains made by the production and domestic sectors (15). If this trend continues, carbon dioxide emissions from transport may exceed those from industry in 1995!

There are currently some 20 million cars on British roads, some 80% of total road traffic. Mean traffic forecasts project increases of motor vehicles on British roads, as shown in Table 1.5 (15).

The number of vehicles on UK roads has increased at about 5% per annum since 1950. Projecting this forward, this number may double the 1990 figure by the year 2004. As Table 1.5 indicates, the Department of Transport predicts half this rate of increase, doubling vehicles at around 2020. At that time energy used in transport might be 66% of the total energy delivered to final users.

The rates of increase of VOCs, NOxs, CO and CO_2 from road vehicles have been of the order of 2–3% per annum due to improvements in vehicle efficiencies and the reduction of exhaust emissions. Nevertheless, future projections show that these emissions will double at some time between 2014 and 2025.

Energy and environmental costs of manufacturing systems and products

Energy is used and pollution produced in the exploration, extraction, supply, conversion and utilisation of fossil fuels and raw materials, in transportation, and in the manufacture, advertising, marketing, distribution, retailing and utilisation and disposal of products. Conservation-conscious planning and design can reduce energy and materials consumption and concomitant environmental pollution in each of these activities.

Environmental auditing of products should adopt a 'cradle-to-grave' approach. Environmental ratings and comparisons among products should consider the energy and raw materials requirements and the air, water and soil pollution produced by a product in production, distribution, utilisation and ultimate disposal to the environment. The environmental impact of a product should begin at the design stage, commencing with the question: *Is this product really necessary?* 'Green' consumerism, acting locally in response to perceived global problems, will possibly reject items considered to be unnecessary or undesirable.

Toxic materials, unnecessary packaging, disposable, unnecessary or dangerous products, irreparable items, energy-intensive products and systems and environmentally polluting items (in manufacture, use or disposal) should all be avoided.

Some of the various items in these categories include clingfilm, aluminium foil, food processors, electric carving knifes, electric can openers, sandwich toasters, coffee machines, cosmetics, shower gels, bath foams and oils, hair gels and sprays, home perm kits, plastic furniture, foamed upholstery, dishwashers, detergents and washing powders, bleaches, dry cleaning fluids, cleaning liquids, fly and wasp killers, air fresheners, toilet blocks, descalers, toilet cleaners, bath foams, power tools, batteries, electric lawn mowers, strimmers, shredders, synthetic fertilisers, artificial insecticides, weed controllers, insect repellents, etc., etc.—the list is endless! (18,19).

'Throwaway' society

Total waste arisings in the UK are estimated (15) to be around 400 million tonnes per annum, broken down as shown in Table 1.6.

Although household and commercial waste represent only 9% of total waste arisings, these sectors account for 36 million tonnes of rubbish which must be collected and disposed of, mainly to landfill.

Over 15% by weight and 25% by value of all products purchased in the UK are designed to be thrown away almost immediately. 25% of the plastic produced and 50% of the paper produced is used for packaging, newspapers and plastic cups and cutlery.

Table 1.6 Breakdown of waste arisings in the UK

Mining and quarrying waste	27%
Agricultural waste	20%
Industrial waste	17%
Dredged materials	11%
Waste from demolition and construction	8%
Sewage sludge	8%
Household waste	5%
Commercial waste	4%

Reuse and recycling

It is much more cost-effective to reuse a product rather than to recycle its constituent materials. Metals, toxic chemicals, paint, sump oil, organic materials, bottles and paper should thus be separated out prior to disposal.

It has been estimated that 50% of domestic refuse might be recoverable, worth £2 billion/annum. This would involve waste sorting and collection, shredding organic materials to produce compost, recycling paper, glass, cans and waste oils.

Products should be:

Built to last
Built simple
Built modular

Almost every discarded material can be reclaimed, reused, recycled or combusted to produce heat, work, fuels, chemicals or materials. There exist no technical barriers to waste recovery, but the recovery of energy and materials requires the expenditure of energy and materials. Thus recovery projects, albeit stimulated by concerns for environmental conservation, are motivated by economics and social factors.

Chapter 6 deals with waste and waste management in further detail.

1.5 SUSTAINABLE DEVELOPMENT

> Development which meets the needs of the present generation without compromising the ability of future generations to meet their own needs. (20)

An Indian phrase says that:

> We have not inherited the Earth from our parents
> We have borrowed it from our children.

The concept of sustainable development was central to the UN Earth Summit held at Rio de Janeiro in June 1992. Agenda 21 emerged from this conference, calling for governments to prepare national strategies for sustainable development.

Global environmental burden (21,22)

Global environmental burden = Global population × GNP/capita
× environmental impact/unit GNP
(gross national product)
Global population (1991) = 5.3 billion and will double in 40 years (23)
GNP/capita should increase by 5% per year, rising by a factor of 5 within 40 years
Present global environmental burden = 5.3 billion × 1 × 1 = 5.3 billion
Global environmental burden in 40 years' time
= 10.6 billion × 5 × environmental impact/unit GNP in 40 years' time
Solving,
necessary environmental impact/unit GNP in 40 years' time = 5/50 = 10%.

Thus, to maintain the current level of global environmental burden, environmental impact/unit GNP must fall by 90% within 40 years. Industry must therefore aim to target a 90% reduction in resource utilisation and environmental pollution by the year 2031. This corresponds to a sustained reduction in global resource consumption and environmental pollution per unit of GNP of 5.5% per annum, starting in 1991!

In order to achieve a renewable *energy* economy without incurring great social strife, world energy consumption will have to be reduced by 1% per annum indefinitely (2).

The widespread application of sensible and systematic integrated environmental management procedures and practices, as outlined in this book, will allow the necessary reductions to be made in the short term. In the medium term, consumer attitudes towards consumerism versus conservationism are beginning to swing to the favour of environmental protection.

In the long term, social changes with respect to transportation systems and the work ethic will ensure renewable resource economies, where nations, having greatly reduced demands for energy and material resources, are sustained by renewable energy technologies: biomass, solar, wind, waves, tides and hydroelectric power.

1.6 NATIONAL RESPONSES

EC Directives

In response to local and global environmental problems, various Directives relating to environmental protection have been issued by the European Commission over the past few years. Because the Commission has little direct control over national legislation, individual member states are expected to act according to these

Directives within the legislative frameworks of the individual states under the terms of subsidiarity.

The Fifth European Community Environmental Action Programme

This programme focuses upon the improved enforcement of existing legislation, but also indicates the probable development of EU environmental policies, leading to further new legislation over the coming years. Until recently, this policy has been aimed at controlling and reducing the environmental impact of industry (i.e. end of pipe clean-up (Figure 1.4)). The next steps (after 1996) involve the adoption of clean technologies (e.g. with regard to packaging, landfills, incinerators, etc.) and the development of environmentally sound products. In other words, EC directives, national legislation and standards will start to creep up the two-pipe system.

The consumer will be told:

Thou shalt not discard . . .
Thou shalt not accept products which . . .

The producer will be told:

Thou shalt not use . . .
Thou shalt not produce . . .

These will be accompanied by the requirement for public information, the introduction of levies (e.g. carbon taxes) based upon the 'polluter pays' principle, incentive schemes and increased civil liabilities.

UK policy

In the UK for example, the government White Paper, *This Common Inheritance* (24) was issued in September 1990. This represented the UK's first comprehensive survey of all aspects of local and global environmental concern and included considerations of the following.

Global environmental problems

- Tropical rainforest destruction
- Stratospheric ozone depletion
- Loss of plant and animal species (biodiversity)
- Drought
- Desertification

- Population growth
- Global warming

Local environmental problems

- Acid rain
- Vehicle pollution
- Industrial pollution
- Water pollution
- Waste disposal

It was seen that there are two broad approaches to control pollution and to tackle environmental problems:

- To introduce regulation by which standards or actions are applied by law.
- To impose market regulators through which taxes and prices could be used to influence producers and consumers.

To combat the perceived environmental problems, therefore, the following recommendations were made.

Energy and air pollution

- To encourage the more prudent and efficient use of energy
- To promote combined heat and power schemes
- To improve guidance to motorists on how to save fuel
- To consider changes to fuel and energy taxes
- To extend the MOT test to cover emissions
- To encourage the use of public transport, when appropriate
- To study where land use planning might help reduce demands for travel
- To extend and improve the monitoring of air pollution
- To maintain strong controls over waste and pollution
- To ensure that power generation companies fit **flue gas desulphurisation** equipment where needed
- To ensure that power generation companies install NOx burners where needed
- To press for European Union action:
  - to adopt minimum standards for appliances and equipment via an **eco-labelling** scheme and to encourage this at home
  - to adopt a new system of **integrated pollution control**
  - to finalise tight emission standards, testing and enforcement to control smoke from heavy diesels and all new cars
  - for new lorry standards to reduce NOx emissions

— to work to control the volatile organic compounds (VOCs) which contribute to the production of ground-level ozone and to control ozone levels
— for new directives on toxic and dangerous waste incineration

Water and pollution

- To introduce regulations setting minimum standards for the construction of silage and slurry stores and to review the need for more control over livestock waste
- To work to:
  - cut inputs of dangerous substances to water courses
  - stop industrial waste disposal and the dumping of sewage sludge at sea
- To maintain the right organisations to supply water, manage it and protect its quality
- To set national water quality standards
- To ensure that the activities of water polluters are regulated and restricted
- To work for an EC directive on nitrate control

Waste and pollution

The UK government aims to seek the best use of materials and the safe and efficient disposal of waste by:

- encouraging:
  - waste minimisation
  - materials recycling
  - energy recovery
- imposing tight controls over waste disposal

and has set a target to recycle 50% of the domestic waste which can be recycled by:

- working with local authorities to assess the effectivenesses of recycling projects
- encouraging more recycling 'banks'
- working up a scheme for labelling recycled and recyclable products
- looking for ways to encourage companies to recycle and to extract recyclable products

Grants have been made available towards 'clean technologies' which help to minimise waste products and energy use.

UK legislation and regulations

The relevant legislation covers, in general, **energy**, **water** and **waste**.

Environmental law affects industry in three ways:

(1) Certain activities must be authorised by a regulatory agency before they are undertaken. Action without the requisite authorisation is usually a criminal offence and may receive penalties of fines and/or imprisonment.
(2) Duties are imposed upon industrial operators. Failure to fulfil these duties is usually a criminal offence with the above penalties.
(3) Legal liabilities for damage, etc. may be imposed.

In 1987, **Her Majesty's Inspectorate of Pollution (HMIP)** was created in order to provide more cost-effective control of major industrial processes.

The **UK Environmental Protection Act 1990** (25)

- tightened waste disposal standards through a *duty of care* on all those who handle waste, with respect to safe treatment and the prevention of losses. Breach of the duty of care constitutes a criminal offence.
- separated the roles of local authorities so that waste disposal operators are no longer responsible for checking their own standards. Thus the functions of **Waste Regulation Authorities** and **Waste Disposal Authorities** were separated.
- enabled those who save disposal costs by *recycling* to earn financial credits.
- introduced **integrated pollution control**

The **Environmental Protection (Prescribed Processes and Substances) Regulations 1991** (26) list the processes for which an authorisation from HMIP is required (see Chapter 2). HMIP may then set conditions for emissions from the process to air, land and water. One objective is that the best available techniques not entailing excessive costs (**BATNEEC**) are used to prevent or minimise pollutant releases to the environment.

Also established under the Environmental Protection Act is Local Authority Air Pollution Control (**LAAPC**), which regulates lighter industrial processes which emit pollutants to the air.

Statutory nuisances are defined as accumulations, deposits, fumes, smells, etc. which constitute a nuisance or which are prejudicial to health. Local authorities may serve abatement and clean-up notices on persons responsible for the nuisance. Failure to comply with such a notice constitutes a criminal offence. Best practical means (**BPMs**) should be taken to prevent the nuisance occurring.

A duty was imposed upon local authorities to inspect land for accumulations of noxious gases and liquids caused by the deposit of controlled waste. Where such concentrations are found, the waste regulation authority must clean up the area in question and may recover the costs involved from the owner of the land.

In 1989, the UK **Water Act** (27) created the **National Rivers Authority (NRA)** as a water pollution 'watchdog'. This authority controls pollution of rivers, estuaries and bathing waters and manages water resources. It protects against floods, supervises fisheries, nature conservation and recreation for inland waters. Privatised water companies with responsibility for the quality of drinking water, rivers and bathing waters were also created and maximum fines for water pollution offences were increased.

In 1991, the **Water Resources Act** (28) introduced substantial provisions for public registers to list consents authorised by the NRA to discharge into controlled waters. The NRA was also given the power to clean up contaminated controlled waters and then recover the costs involved from the polluter.

The **Water Industry Act** 1991 (29) requires that consent be obtained from a water company to discharge trade effluent to sewers.

Penalties

The consequences of pollution offences can be

- criminal liability
- civil liability
- liability for clean-up costs

The maximum fine in magistrates' courts for most offences is now £20 000. If the prosecuting authority (e.g. HMIP or NRA) brings proceedings in the Crown Court, then the fine is unlimited. Civil liability with respect to damage caused is also unlimited and may result in an injunction being granted against a polluting company. Failure to comply with a court order can result, on conviction or indictment, in imprisonment for a term not exceeding two years.

Liability can now be imposed upon company directors and others at a personal level. The Act lists directors, managers, secretaries and similar officers purporting to act for the company and where there is consent, connivance or neglect.

Legislation and regulations are covered in more detail in Chapter 2.

Standards

The British Standards Institution (**BSI**) has recently produced a specification for **Environmental Management Systems—BS 7750:1992** (30), similar to BS 5750:1987 (31, which deals with quality systems) with which it is analogous, and to which it has many links.

The European Commission has also produced a draft Eco-Audit Regulation (32) which follows similar lines. BS 7750 also links to this regulation.

BS 7750 contains a specification for an environmental management system for ensuring and demonstrating compliance with stated environmental management policies and objectives. It is designed to enable an organisation to establish an effective environmental management system, as detailed in Chapter 2.

1.7 CONSIDERATIONS FOR COMPANIES

Seven stages of a project

- wild enthusiasm
- frantic activity
- realisation and disillusionment
- search for the guilty
- punishment of the innocent
- promotion of the non-participants
- definition of the problem

Example of interpretation

• wild enthusiasm	global warming, ozone layer depletion
• frantic activity	effects—carbon taxes, VAT, abolish CFCs
• realisation and disillusionment	costs and diseconomies, seeds of doubt
• search for the guilty	scientists and politicians
• promotion of the non-participants	new leaders
• definition of the problem	local pollution

Why should a company seek to protect the environment?

- **To comply with environmental legislation and so avoid fines and other penalties.** Legal liability for environmental damage dictates that any party that causes environmental damage may be fined and required to bear the costs of remediation and make financial compensation for any losses endured by third parties. Thus banks and other lenders, as well as shareholders and insurance companies, are insisting upon high environmental standards as a condition attached to loans or insurance policies.
- **To conserve resources by using less input materials and energy and so saving money.** There is no doubt that the costs of energy, electricity, water supplies and raw materials will rise substantially in real terms, as the supply companies comply with new environmental legislation and standards. The use of less input resources results in less emissions to air and water and less solid waste. This leads to lower levels of pollution taxes, effluent charges and waste disposal costs. These costs will also rise substantially over the coming years as the effluent and waste disposal operators also have to comply with high environmental standards and legislation.
- **To adopt high environmental standards** by obtaining accreditation to BS 7750, the British Standard on Environmental Management Systems, and to comply with the requirements of the EC Eco-management and Audit Scheme.

The above are necessary because they will (23):

- **Benefit public relations**, enhance competitiveness and gain or retain customers for the company's products.
- **Satisfy suppliers and trading partners**. The knock-on effect in supply chains is already resulting in situations where retail outlets are insisting on their suppliers conforming to recognised environmental standards. This is increasingly being written into contracts.
- **Demonstrate to the local community** that it is not being exposed to environmental risk due to the company's operations.
- **Communicate the company's environmental policy to employees**, so that the company may attract, retain and motivate a quality workforce.
- **Keep the media at bay**

What must a company do?

- Conserve resources and so save money whilst protecting the environment.
- Avoid legal action, fines and pollution taxes.
- Be seen to be environmentally responsible and so gain a competitive advantage.

REFERENCES

(1) HMSO, *Greenhouse Effect, Report of House of Lords Select Committee on Science and Technology, Volume II—Evidence*, HMSO, 1989.
(2) P.W. O'Callaghan, *Energy Management*, McGraw-Hill, London, 1993.
(3) M. McElroy, The Challenge of Global Change, *New Scientist*, 34–36, 28 July 1988.
(4) P. Smith, RIBA Conference Discusses Greenhouse Effect, *Energy Management*, 18–21, UK DEn., April/May 1990.
(5) G. Brookes, The Greenhouse Effect, *ESTA Energy Efficiency Yearbook*, UK 1990.
(6) J. Gribbon, *Hothouse Earth: The Greenhouse Effect and Gaia*, Bantam Press, London, 1990.
(7) L.R. Embar, P.L. Layman, W. Lepkowski and P.S. Zurer, Tending Global Commons—The Changing Atmosphere, *Chemical and Engineering News*, **64**, 47, 16–35, November 1986.
(8) R.A. Houghton and G.M. Woodwell, Global Climatic Change, *Scientific American*, **260**, 4, 18–26, April 1989.
(9) E.A. Avallone and T. Baumeister III, *Marks' Standard Handbook for Mechanical Engineers*, 9th Edition, McGraw-Hill, New York, 1987.
(10) P.H. Collin, *Dictionary of Ecology and the Environment*, Peter Collin Publishing, 1988.
(11) A. Porteous, *Dictionary of Environmental Science and Technology*, Open University Press, Milton Keynes, 1991.
(12) *The Guardian*, 23 February 1990.
(13) A.E.B. Taylor, Energy and Environmental Policies for Agriculture, MSc Thesis, Cranfield Institute of Technology, 1991.
(14) M. Allaby, *The Survival Handbook*, Pan UK, 1975.

(15) A. Brown, *The U.K. Environment*, Department of the Environment, Government Statistical Service, HMSO, London, 1992.
(16) Department of Energy, *An Evaluation of Energy Related Greenhouse Gas Emissions and Measures to Ameliorate Them*, Energy Paper Number 58, UK Country Study for the Intergovernmental Study on Climate Change Response Strategies Working Group, Energy and Industry Sub Group, HMSO, October 1989.
(17) Department of the Environment, *U.K. Strategy for Sustainable Development*, Consultation Paper, July 1993.
(18) J. Elkington and J. Hailes, *The Green Consumer Guide*, Gollancz, London, 1989.
(19) J. Elkington and J. Hailes, *The Green Consumer's Supermarket Shopping Guide*, Gollancz, London, 1989.
(20) The Bruntland Report, World Commission on Environment and Development, *Our Common Future*, Oxford University Press, Oxford, 1987.
(21) P. Ehrlich and A. Ehrlich, *The Population Explosion*, Arrow, London, 1991.
(22) A. Stikker, Sustainability and Business Management, *Business Strategy and the Environment*, 1, 3, Autumn, 1992.
(23) R. Welford and A. Gouldson, *Environmental Management and Business Strategy*, Pitman Publishing, London, 1993.
(24) *This Common Inheritance*, HMSO, 1990.
(25) *Environmental Protection Act*, HMSO, 1990.
(26) *Environmental Protection (Prescribed Processes and Substances) Regulations*, HMSO, 1991.
(27) *Water Act*, HMSO, 1989.
(28) *Water Resources Act*, HMSO, 1991.
(29) *Water Industry Act*, HMSO, 1991.
(30) British Standards Institution, *Specification for Environmental Management Systems, BS 7750*, 1992.
(31) British Standards Institution, *Quality Systems, BS 5750*, 1987.
(32) *EC Eco-Audit Regulation* (Draft Version 3, December 1991).

2

Environmental legislation: what are the requirements of the law?

This chapter deals with EC and UK legislation with respect to environmental pollution. It reviews the concepts of nuisance and duty of care in this regard and then examines environmentally related EC Action Programmes and Directives. The major EC Directives, UK Acts and Regulations concerning air pollution, water pollution and waste on land are listed chronologically. The UK Environmental Protection Act 1990 is considered in some detail. The concept and aims of integrated pollution control are defined and the principle of BATNEEC described. An Appendix contains the list of prescribed processes and substances and examples of BATNEEC for the reduction of emissions are collated. The application procedure for HMIP IPC authorisation to emit is also included.

Acronyms

BATNEEC	best available techniques not entailing excessive cost
BPM	best practicable means (displaced by BATNEEC in the EPA)
BPEO	best practicable environmental option
BS	British Standard
BSI	British Standards Institution
EC	European Commission
EPA	Environmental Protection Act 1990
HMIP	Her Majesty's Inspectorate of Pollution
HMSO	Her Majesty's Stationery Office
IPC	integrated pollution control
IPPC	integrated pollution protection and control
LAAPC	Local Authority air pollution control
LAWDC	Local Authority Waste Disposal Companies
NRA	National Rivers Authority

WCA	Waste Collection Authorities
WDA	Waste Disposal Authorities (became WRAs in the EPA)
WRA	Waste Regulation Authorities

2.1 ENVIRONMENTAL LAW

The relevant legislation covers, in general, energy, water and waste, in relation to pollution via emissions to air, effluent discharges to water and the disposal of waste to land.

Environmental law affects industry in three ways:

- Certain activities must be authorised by a regulatory agency before they are undertaken. Action without the requisite authorisation is usually a criminal offence and may receive penalties of fines and/or imprisonment.
- Duties are imposed upon industrial operators. Failure to fulfil these duties is usually a criminal offence with the above penalties.
- Legal liabilities for damage, etc., may be imposed.

Nuisance and duty of care

Public nuisance is a tort (a civil wrong), being an act or omission which materially affects the material comfort and quality of life of the public generally or of an identifiable sector of the public.

Private nuisance, affecting an individual or a group of individuals, is also a tort.

Statutory nuisances are those specific nuisances at common law that have been included within statute.

Environmental statutory nuisances are defined within the Environmental Protection Act (EPA) 1990 as *accumulations, deposits, fumes, smells, etc. which constitute a nuisance or which are prejudicial to health.*

Duty of care is defined as *that degree of care which does not cause injury to a neighbour.* In the Environmental Protection Act 1990, the duty of care requires one to take all such measures as are reasonable in the particular circumstances. Reasonable care must be taken to avoid acts or omissions which can be reasonably foreseen as likely to injure a neighbour (*one of those persons who are so closely and directly affected by an act that the person causing the act ought reasonably to have foreseen that they would be so affected when contemplating the act or omission*).

2.2 EC DIRECTIVES

Various environmentally related Directives have been issued by the European Commission over recent years. The results to be achieved by set dates are binding on the member states but, under the principles of subsidiarity, the methods of implementation are left to each state. The **Single European Act 1986** introduced the *co-operation procedure* which allows the European Parliament to express an opinion on the draft of EC legislation and a common position, reached by majority vote. The EC therefore produces *Action Programmes* setting out the agreed programme of activities for coming years. Since 1972, there have been five European Community Action Programmes on the environment. Initially, the essential aim of community environmental policy was the control of pollution and nuisances. Later, awareness concentrated on prevention leading to the integration of environmental requirements into the planning and execution of actions in many economic and social sectors.

The First Environmental Action Programme (1973)

This programme set out to establish environmental quality standards and listed a number of principles which have formed the basis of subsequent programmes. These include the following:

- Prevention is better than cure
- Environmental effects should be taken into account at the earliest stage in decision making
- Exploitation which causes significant damage to the ecological balance must be avoided
- Scientific knowledge must be improved to enable action to be taken
- The 'polluter pays'
- Activities in one member state should not cause environmental damage in another
- The interests of developing countries should be considered
- EC and member states should act together
- Education of citizens is important

The Second Environmental Action Programme (1977)

This updated the first programme in order to ensure the continuity of the projects already being undertaken and to undertake new tasks for the period 1977–81.

The Third Environmental Action Programme (1981)

Until this programme, the essential aim of Community environmental policy was the control of pollution and nuisances. This programme started to concentrate on the prevention of pollution through the integration of environmental requirements into the planning and execution of actions, and introduced the need for *environmental impact assessments* of planned new activities.

The Fourth Environmental Action Programme (1987)

This covered the period from 1987 to 1992, and was based upon the recognition that the Community's environmental protection policy could contribute to improved economic growth and job creation. Its objective was to preserve, protect and improve the quality of the environment, to contribute towards protecting human health, and to ensure a prudent and rational use of natural resources. It:

- gave priority to implementation of existing legislation
- suggested that environmental policies should be integrated into economic, industrial, agricultural and social policies
- considered the use of economic instruments as a means of implementing environmental policies
- stated a need for an EC Freedom of Environmental Information Act
- introduced the concept of *integrated pollution prevention and control (IPPC)*
- suggested that water quality should be improved and soil should be protected
- identified the need for new technologies to reduce waste and improve waste management
- promoted the introduction of *environmentally friendly* goods and services
- called for an inventory of emissions and major sources
- called for an inventory of *best pollution abatement technologies*
- called for the development of new low-pollution technologies

The Fifth Environmental Action Programme (1993)

This had its basis in the principles of *sustainable development*, preventative and precautionary action and shared responsibility. Environmental policy was required to contribute to promoting measures at international level to deal with regional or world wide environmental problems. The word *sustainable*, was intended to reflect a policy and strategy for continued economic and social development without detriment to the environment or the natural resources on the quality of which continued human activity and further development depend. It focused on the agents and activities which deplete natural resources and otherwise damage the environment, rather than to wait for problems to emerge *at the end of the pipe*. Whereas the previous Action Programmes relied almost exclusively on legislative measures, the mix of instruments proposed to achieve the objectives of the fifth Action Programme included legislative, market-based, and financial instruments, involving not only government and manufacturing industries, but also public authorities, public and private enterprises in all their forms, as well as the general public, both as citizens and consumers. Five target sectors were chosen for special attention: **industry**, **energy**, **transport**, **agriculture** and **tourism**, because these sectors have crucial roles in sustainable development.

Three questions to ask when assessing the degree of sustainability of a system are:

- Is the system self-sufficient in energy, water and materials?
- Does the system pollute the environmental media: air, water or land?
- Will the system be operating in exactly the same manner in 100 years' time, if left to its own resources?

Thus the Fifth Environmental Action Programme focused on the improved enforcement of existing legislation, but also indicated the probable development of EC environmental policies, leading to further new legislation over the coming years. Until recently, this policy had been aimed at controlling and reducing the environmental impact of industry (i.e. end of pipe clean-up). The next steps (after 1996) involve the adoption of clean technologies (e.g. with regard to, for example, packaging, landfills, incinerators, etc.) and the development of environmentally sound products. These will be accompanied by the requirement for public information, the introduction of levies (e.g. carbon taxes) based upon the 'polluter pays' principle, incentive schemes, and increased civil liabilities. It was perceived that it had become necessary to strike a new balance between the short-term benefits of individuals or corporate bodies and the long-term aims of society as a whole.

Priority was accorded to the following areas:

- sustainable management of natural resources
- integrated pollution control
- prevention and management of waste
- reduction in the consumption of non-renewable energy
- more efficient management of mobility
- improvement of the urban environment
- improvements in health and safety

Research and development were seen to be needed to provide improved early warning of environmental problems, to improve understanding of the impacts of human activities on basic environmental processes, and to develop and apply techniques for preventing, reducing and mitigating environmental impacts. Specific subject areas for attention included climate change, ozone layer depletion, acidification, deforestation, air quality, the protection of nature and biodiversity, water quantity and quality, noise, coastal zones, municipal waste and hazardous wastes. New instruments to be introduced included *eco-audits*, *ecological labelling*, public information, education and *access to environmental information*.

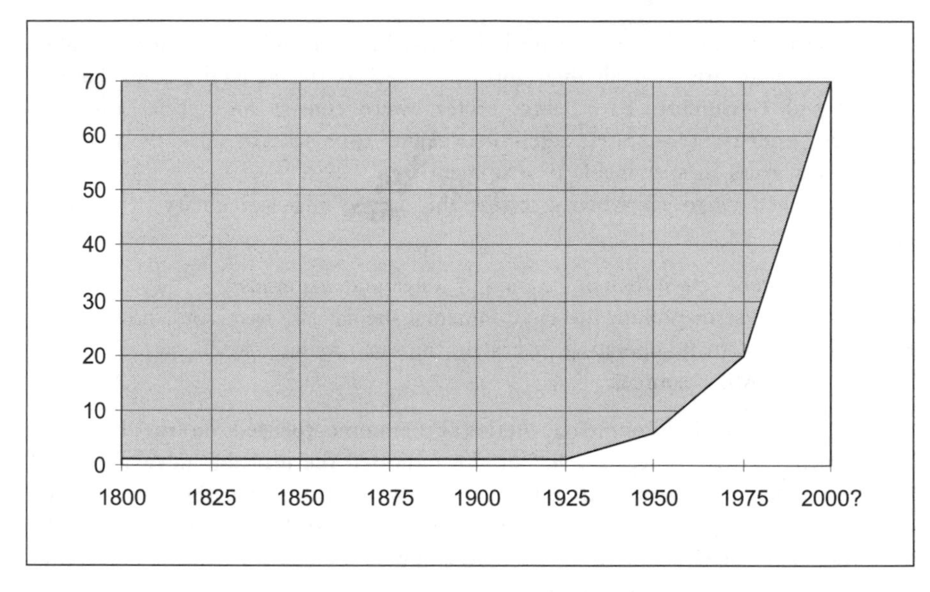

Figure 2.1 Some UK Acts and Regulations relating to pollution of the environment

2.3 CHRONOLOGY

The major existing EC Directives, UK Acts and Regulations are listed in Table 2.1. This list is not intended to be exhaustive, but includes most of the items which might affect companies and organisations in the manufacturing and service sectors.

The number of legislative instruments for the reduction of pollution have increased manyfold in recent years (Figure 2.1).

2.4 AIR POLLUTION

Controls on air pollution from industrial works were first developed in the UK in the 1860s, with the requirement that the *Best Practicable Means (BPMs)* should be applied to control the release of substances harmful to the environment. The **Clean Air Acts, 1956** and **1968** were concerned with smoke, grit and dust from combustion processes and required BPMs to prevent these emissions.

2.5 WATER POLLUTION

The **Water Act 1973** placed a general duty on the Secretary of State to secure the execution of a national policy for water, including the restoration and maintenance

Table 2.1 Some EC Action Programmes and Directives, UK Acts and Regulations relating to the environment

Year	Type	Development	PRO	AIR	WAT	LAN
1876	ACT	Rivers Pollution Prevention Act			×	
1906	ACT	Alkali, etc. Works Regulation Act	×	×	×	×
1936	ACT	Public Health Act				×
1937	ACT	Public Health (Drainage of Trade Premises) Act			×	
1945	ACT	Water Act			×	
1947	ACT	Town and Country Planning Act				×
1951	ACT	Rivers Pollution Prevention Act			×	
1956	ACT	Clean Air Act	×	×		
1960	ACT	Clean Rivers (Estuaries and Tidal Waters) Act			×	
1961	ACT	Rivers (Prevention of Pollution) Act			×	
1961	ACT	Public Health Act			×	
1968	ACT	Clean Air Act	×	×		
1971	ACT	Rural Water Supplies and Sewerage Act			×	
1971	REG	Clean Air (Emission of Grit and Dust from Furnaces) Regulations		×		
1972	ACT	Local Government Act				×
1973	ECAP	First EC Environmental Action Programme	×	×	×	×
1973	ACT	The Water Act			×	
1974	ACT	Control of Pollution Act	×	×	×	×
1974	ACT	Health and Safety at Work, etc. Act	×	×	×	×
1974	DIR	Quality of Fresh Waters Needing Protection or Improvement in Order to Support Fish Life			×	
1975	DIR	EC Framework Directive on Waste				×
1976	DIR	EC Directive on Pollution caused by certain Dangerous Substances Discharged to the Aquatic Environment of the Community			×	
1976	REG	Control of Pollution (Discharges into Sewers) Regulation			×	
1976	DIR	Quality of Bathing Water			×	
1977	ECAP	Second EC Environmental Action Programme	×	×	×	×
1978	DIR	EC Toxic and Dangerous Waste Directive				×
1978	ACT	Refuse Disposal (Amenity) Act				×
1979	DIR	Quality Required of Shellfish Waters			×	
1980	DIR	Protection of Surface and Coastal Water from Dangerous Substances			×	
1980	DIR	EC Directive Relating to the Quality of Water for Human Consumption			×	
1980	DIR	EC Directive on Air Quality Limits and Guide Values for Sulphur Dioxide and Suspended Particulates		×		

continued overleaf

Table 2.1 *(continued)*

Year	Type	Development	PRO	AIR	WAT	LAN
1980	REG	Control of Pollution (Special Waste) Regulations				×
1981	ECAP	Third EC Environmental Action Programme	×	×	×	×
1982	DIR	EC Directive on a Limit Value for Lead in Air		×		
1983	REG	Health and Safety (Emissions into the Atmosphere) Regulations		×		
1984	DIR	Combating of Air Pollution from Industrial Plants Directive	×	×	×	
1984	REG	Control of Pollution (Consents for Discharges) Regulations			×	
1985	ACT	Food and Environmental Protection Act	×			
1985	DIR	EC Directive on Air Quality Standards for Nitrogen Dioxide		×		
1986	ACT	The Single European Act				
1987	ECAP	Fourth EC Environmental Action Programme	×	×	×	×
1988	REG	Collection and Disposal of Waste Regulations				×
1988	DIR	EC Directive on the Limitation of Emissions of Certain Pollutants into the Air from Large Combustion Plant		×		
1988	ACT	Environment and Safety Information Act	×	×	×	×
1989	REG	Surface Water (Classification) Regulations			×	
1989	ACT	Control of Pollution (Amendment) Act	×			×
1989	REG	Air Quality Standards Regulations		×		
1989	ACT	The Water Act			×	
1989	REG	Trade Effluents (Prescribed Processes and Substances) Regulations			×	
1989	ACT	Control of Smoke Pollution Act	×	×		
1989	REG	Control of Industrial Air Pollution (Registration of Works) Regulations	×	×		
1989	REG	Water Supply (Water Quality) Regulations			×	
1989	REG	Control of Pollution (Consents for Discharges) Regulations			×	
1989	REG	Control of Pollution (Registers) Regulations			×	
1990	ACT	Planning (Hazardous Substances) Act	×	×	×	×
1990	ACT	Environmental Protection Act	×	×	×	×
1991	DIR	EC Framework Directive on Waste				×
1991	DIR	EC Directive concerning Urban Waste Water Treatment			×	
1991	REG	The Environmental Protection (Prescribed Processes and Substances) Regulations (replaces 1983 REG)	×	×	×	×
1991	REG	Environmental Protection (Applications Appeals and Registers) Regulations	×	×	×	×

Table 2.1 *(continued)*

Year	Type	Development	PRO	AIR	WAT	LAN
1991	REG	Environmental Protection (Authorisation of Processes) Regulations	×	×	×	×
1991	ACT	Water Resources Act			×	
1991	ACT	Water Industry Act			×	
1991	REG	EC Eco-Audit Regulation	×	×	×	×
1992	REG	EC Ecolabelling Regulation	×	×	×	×
1993	ECAP	Fifth EC Environmental Action Programme	×	×	×	×

Notes:
ACT—UK Act
REG—UK or EC Regulations
DIR—EC Directive
ECAP—EC Action Programme
PRO—processes
AIR—emissions to air
WAT—effluents to water
LAN—waste on land

of the wholesomeness of rivers and their inland waters. The **Control of Pollution Act 1974** required that discharges to rivers, specified underground waters or land must have the *consent* of the **Water Authority**.

The EC Directive on **Pollution Caused by Certain Dangerous Substances Discharged to the Aquatic Environment of the Community 1976** gave a black list of 129 toxic substances, e.g. mercury, cadmium, carcinogens, lindane, DDT, carbon tetrachloride, entachlorophenol, aldrin, dieldrin, from which water pollution *should be eliminated* and a grey list of less toxic substances, e.g. chromium, lead, zinc, copper, nickel and arsenic, from which water pollution *should be reduced by limitation*.

The **Trade Effluents (Prescribed Processes and Substances) Regulations 1989** listed the most dangerous 'red list' substances, which became the SCHEDULE 5 substances, listed in the **Environmental Protection (Prescribed Processes and Substances) Regulations 1991** (contained in Appendix 2.1).

The **Water Act 1989** introduced a system of statutory water quality objectives and standards for drinking water. It created the **National Rivers Authority (NRA)** as a water pollution 'watch dog', having responsibility for the monitoring and control of water pollution. This Authority controls pollution of rivers, estuaries and bathing waters and manages water resources. It protects against floods, and supervises fisheries, nature conservation and recreation for inland waters. The NRA exercises statutory powers for controlling prescribed substances and emissions to water from prescribed processes and audits reports on the activities of the **Waste Regulatory Authorities (WRAs)**. Consent from the NRA is required for discharges to inland waters. It became a criminal offence to discharge any poisonous, noxious or polluting matter to any stream or controlled water without this consent. The setting up of the NRA paved the way for the privatised **water companies**, separating the roles of discharger and regulator. These water companies have responsibilities for the quality of drinking water, rivers and bathing waters. Maximum fines for water pollution offences were increased.

The **Drinking Water Inspectorate (Water Supply (Water Quality) Regulations 1989)** ensures that water companies fulfil statutory requirements for the supply of quality drinking water. The **Water Resources Act 1991** introduced substantial provisions for public registers to list consents authorised by the NRA to discharge into controlled waters. The NRA was also given the power to clean up contaminated controlled waters and then recover the costs involved from the polluter. The **Water Industry Act 1991** required that consent be obtained from a water company to discharge trade effluent to sewers.

Thus all liquid emissions are controlled either by the water companies or by the National Rivers Authority.

The **Environmental Protection Act 1990** required that integrated pollution control (IPC) authorisation from HMIP be obtained for those most polluting plants which release 'red list' substances into controlled waters or sewers.

The **EC Directive concerning Urban Waste Water Treatment 1991** indicated that it would become unlawful to discharge industrial waste into the municipal system without prior *authorisation* from the relevant authority. It requires that certain standards be applied in terms of treatment in certain industries.

2.6 WASTE ON LAND

The **Local Government Act 1972** introduced the concept of **Waste Collection Authorities (WCA)** and **Waste Disposal Authorities (WDA)**. The **Control of Pollution Act 1974** introduced major requirements for the treatment of waste on land, including the need for *licensing* sites, storage systems, treatment processes and disposal routes for controlled waste, defined as household, industrial and commercial waste (**Collection and Disposal of Waste Regulations 1988**).

The European Community's policies on waste were established in the **EC Framework Directive on Waste 1975**, which required that waste disposal does not present a risk to human health or the environment. This Directive was superseded by the **EC Framework Directive on Waste 1991** which encouraged:

- waste minimisation through
  - avoidance
  - reduction
  - reuse
  - recycling
- the adoption of clean technologies
- national and EU self-sufficiency in waste management (i.e. waste should not cross national or EU boundaries)

This Framework Directive also included Directives on

- Integrated Pollution Prevention and Control (IPPC)
- Hazardous Waste and its Incineration
- Liability for Damage Caused by Waste
- Recycling of Packaging Wastes
- Freedom of Access to Environmental Information

The UK **Environmental Protection Act 1990** defined waste as *any substance which constitutes a scrap material or an effluent, or other unwanted surplus substance arising from the application of a process and any substance or article which requires to be disposed of as being broken, worn out, contaminated or otherwise spoiled.* It classified waste as:

- *controlled waste*—household, commercial and industrial solid waste (not explosives, agricultural or mining waste)
- *special waste*—which is dangerous or difficult to handle, transport or treat
- *hazardous waste*—which relates to special waste often transported across national boundaries

Other definitions of waste include:

- *municipal waste*—which is collected by Waste Collection Authorities (WCAs)
- *inert waste*—which does not decompose, or decomposes very slowly
- *semi-inert waste*—which may decompose slowly, but is only slightly soluble in water
- *putrescible waste*—which may decompose and may consist in part of soluble matter which could cause pollution if allowed to enter ground or surface water systems
- *difficult waste*—which is not special waste, but is difficult to handle or dispose of (e.g. liquids, sludges); it may consist, in part, of soluble matter which would cause pollution if allowed to enter ground or surface water systems
- *civic amenity waste*—from households
- *notifiable wastes*—which are poisonous, noxious or polluting
- *prohibited wastes*—peroxides, chlorates, perchlorates, azides, explosive materials and radioactive materials

The **Control of Pollution (Special Waste) Regulations 1980** dealt with the most hazardous special wastes, seen to be dangerous to life. A *consignment system* covering transportation, packaging and labelling was instituted. These special wastes are:

- acids and alkalis
- antimony and its compounds
- arsenic compounds

- asbestos (all chemical forms)
- barium compounds
- beryllium and its compounds
- biocides and phytopharmaceutical compounds
- boron compounds
- cadmium and its compounds
- copper compounds
- heterocyclic organic compounds containing oxygen, nitrogen or sulphur
- hexavalent chromium compounds
- hydrocarbons and their oxygen, nitrogen and sulphur compounds
- inorganic cyanides
- inorganic halogen-containing compounds
- inorganic sulphur-containing compounds
- laboratory chemicals
- lead compounds
- mercury compounds
- nickel and its compounds
- organic halogen compounds, excluding inert polymeric materials
- peroxides, chlorates, perchlorates and azides
- pharmaceutical and veterinary compounds
- phosphorus and its compounds
- selenium and its compounds
- silver compounds
- tarry materials from refining and tar residues from distilling
- tellurium and its compounds
- thallium and its compounds
- vanadium and its compounds
- zinc compounds

Other special waste is that waste which has:

- the possibility of presenting the risk of death or serious tissue damage
- explosivity
- flammability
- a tendency to produce a flammable gas on contact with air or water
- oxidisability
- acute toxicity
- infectivity
- corrosivity
- reactivity
- delayed toxicity
- ecotoxicity

2.7 HER MAJESTY'S INSPECTORATE OF POLLUTION

In 1987, **Her Majesty's Inspectorate of Pollution (HMIP)** (originally founded in the **Alkali, etc. Works Regulation 1906)** was created in order to provide more cost-effective control of major industrial processes. HMIP deals with the authorisation and oversight of those processes with the greatest potential for pollution, and is charged with the pursuance of *integrated pollution control.* HMIP exercises statutory powers for controlling prescribed substances and emissions to air from prescribed processes, monitors Water and Waste Disposal Authorities and develops *best practical environmental options (BPEOs).*

2.8 ENVIRONMENTAL PROTECTION ACT 1990

The **Environmental Protection Act 1990** drew together most of the environmentally related statutory nuisances from earlier legislation, including those due to:

- premises which are health risks or nuisances
- smoke, fumes, vapours or gases, dust, steam or smells, which are health risks or nuisances
- accumulations or deposits which are health risks or nuisances

It placed a duty of care on producers, importers, holders, carriers of waste and persons who treat waste:

- to take all such measures to avoid waste pollution as are reasonable in the particular circumstances
- to have the responsibility for ensuring that controlled waste is not illegally managed, does not escape from control, is transferred only to an authorised person and is adequately described and clearly labelled to enable proper handling and treatment

The waste disposers are therefore responsible for the safe passage of controlled waste through the waste disposal chain to its *ultimate* safe disposal. Each party in this chain must ensure that *all reasonable measures* are taken to prevent any *unauthorised* or harmful deposit, treatment or disposal of the waste concerned.

Integrated pollution control (IPC) is *the simultaneous protection of air water and land from the release of prescribed substances from prescribed processes*, under the control of HMIP, *to ensure that the best practicable environmental options (BPEO) are adopted*, (i.e. the releases from these processes are controlled through the use of *BATNEEC (the best available techniques not entailing excessive costs)*, so as to have the least effect on the environment as a whole, so that, for example, airborne waste cannot be converted into a liquid waste).

A single written authorisation for a particular process, covering all

three environmental media (air, water and land), with a view to minimising pollution to the environment as a whole, must be applied for, and this will be granted providing that the BATNEEC have been employed.

The Environmental Protection Act (EPA) 1990 is divided into two parts:

- EPA Part 1: Integrated Pollution Control (IPC) and Air Pollution Control by Local Authorities (LAAPC)
- EPA Part 2: Waste on Land

EPA Part 1: Integrated Pollution Control (IPC) and Air Pollution Control by Local Authorities (LAAPC)

Part 1 of the EPA divides industrial and other emitters into Part A and Part B processes.

Part A processes are the most significant polluting processes in the UK and all of their emissions are controlled under the EPA by Her Majesty's Inspectorate of Pollution, HMIP, in England and Wales. These processes are obliged to apply the objectives of integrated pollution control (IPC), whereby the total emissions to air, water and land are considered under a single application for authorisation to emit. The company is required to minimise its overall environmental impact by ensuring that, and convincing HMIP that, the best practicable environmental options (BPEOs) are achieved.

Also established under the EPA is Local Authority air pollution control (LAAPC), which regulates lighter industrial processes which emit pollutants to the air. Thus Part B processes, although more common, are less significant than Part A processes and come under the control of Local Authorities. Although the latter are only obliged to regulate emissions to air, the processes involved are still bound to apply the principles of BATNEEC.

The EPA details all processes which are subject to regulation and control in the UK. All prescribed processes, as defined by the Act, must apply for authorisation to the relevant regulatory authorities in order to emit into the environment.

The **Environmental Protection (Prescribed Processes and Substances) Regulations 1991** list the processes (Appendix 2.1) for which an authorisation from HMIP or Local Authorities is required.

HMIP may then set conditions for emissions from the process to air, land and water. These conditions may relate to the method of operation, the training of staff and abatement techniques to be used to reduce the release of substances. Releases will be regulated by conditions in authorisations limiting explicitly the substances that can be released to the various environmental media, both in terms of their concentrations and the amounts of the substances released (1).

If authorisation to emit is to be granted, the emitter must satisfy the relevant authority that the process for which they are responsible applies the BATNEEC in the control of emissions. HMIP decides what are BATNEEC for an individual process, taking into account variable factors such as configuration, size, other individual characteristics of the process and local environmental factors. Guidance on what constitutes BATNEEC for each description of prescribed process is provided in the Chief Inspector's Guidance Notes. These process-specific guidance notes all have a similar structure:

- introduction and definition of the process
- general requirements for new and existing plant
- release limits to air, water and land, which HMIP believes can be achieved by the application of BATNEEC and BPEO to the process

Annexes cover:

- a description of the process and the plant used
- a list of the prescribed substances most likely to be present in releases to the environment
- the techniques for pollution abatement which represent BATNEEC for the process
- the monitoring necessary to demonstrate compliance with release limits
- additional requirements such as training and contingency planning

EPA Part 2: Waste on Land

The Waste Collection Authorities (WCA) were given responsibility for collecting waste. The Waste Regulatory Authorities (WRA) were given responsibility for dealing with licensing, inspection and regulation of waste disposal sites and the enforcement of legislation. The Local Authority Waste Disposal Companies (LAWDC) were given responsibility for the operation of waste disposal sites.

Local authorities may serve abatement and clean-up notices on persons responsible for statutory nuisances. Failure to comply with such a notice constitutes a criminal offence. A duty was imposed upon local authorities to inspect land for accumulations of noxious gases and liquids caused by the deposit of controlled waste. Where such concentrations are found, the Waste Regulation Authority (WRA) must clean up the area in question and may recover the costs involved from the owner of the land.

Thus the Environmental Protection Act 1990:

- tightened waste disposal standards through a duty of care on all those who handle waste, with respect to safe treatment and the prevention of losses. Breach of the duty of care constitutes a criminal offence. Companies and managers within them are liable to fines and prison sentences should they fail to fulfil their legal obligations in the waste disposal chain;
- imposed civil liability in respect of any person who deposits, causes or permits unlicensed waste disposal or if the waste for which they are responsible causes damage to the environment or to human health;

- separated the roles of Local Authorities so that waste disposal operators were no longer responsible for checking their own standards. Thus the functions of Waste Regulations Authorities (WRA) and Waste Disposal Authorities (WDA) were separated;
- enabled those who save disposal costs by recycling to earn financial credits.

2.9 PENALTIES

The consequences of pollution offences can be:

- criminal liability
- civil liability
- liability for clean-up costs
- liability for damage caused by waste

The consequences of non-compliance under the EPA are that:

- the regulatory authorities can serve enforcement notices if regulations are breached
- the enforcement notice will detail the steps needed to rectify the situation and the period available
- if there is a serious environmental risk, a prohibition notice may be served to close the facility
- continued non-compliance can lead to prosecution, fines and imprisonment

The maximum fine in magistrates' courts for most offences is now £20 000. If the prosecuting authority (e.g. HMIP or NRA) brings proceedings in the Crown Court, then the fine is unlimited. Civil liability with respect to damage caused is also unlimited and may result in an injunction being granted against a polluting company. Failure to comply with a court order can result, on conviction on indictment, in imprisonment for a term not exceeding two years.

Liability can now be imposed upon company directors and others at a personal level. The EPA covers directors, managers, secretaries and similar officers purporting to act for the company where there is consent, connivance or neglect.

IPC has been introduced progressively between 1 April 1991 and 1 November 1995. The overall effect of this new legislation will be an inevitable rise in the costs of effluent and waste disposal.

2.10 APPLICATION FOR IPC AUTHORISATION

Integrated Pollution Control: A Practical Guide (see Bibliography) contains detailed flow charts for applications procedures for IPC authorisation. Application forms for authorisation under IPC and guidance for applicants for authorisation are available through HMSO.

The information required, as specified in the Regulations, is as follows:

- Name, address and telephone number of the applicant
- Address where the prescribed process will be carried out
- Name of the Local Authority
- Outline map or plan
- Description of the proposed process, its physical characteristics, flows of material within it and information related to BATNEEC
- Descriptions of the nature of the product to be produced, the quantities of production and the time of that production
- Details of any neighbouring processes which could affect, or be affected by, the operation of the plant
- An inventory of prescribed substances, or other substances which might be harmful if released, used in connection with, or resulting from, the process
- Information regarding the quantity and nature of releases, and whether or not the raw materials are used in the process
- A description of the techniques to be used (BATNEEC) to prevent or minimise releases into the environmental media, or render harmless any substances as described above
- Details of any proposed release of prescribed substances
- An assessment of environmental consequences
- Proposals for monitoring releases
- Information on how waste is to be disposed of
- Justification that the plant and its proposed activities meet with requirements for BATNEEC and BPEO

The **Environmental Protection (Applications, Appeals and Registers) Regulations 1991** sets out the detailed procedure for applications for IPC authorisation and also places a duty on the applicant to advertise in a local newspaper details of the prescribed process and its location, together with a statement that public representation may be made, and to whom.

2.11 BATNEEC

The principles of IPC state that a process should be designed and operated in such a way that the substances so released have the minimum overall impact on the environment through its three media: air, water and land.

The HMIP series of *Industry Sector Guidance Notes* (IPR 1 to 6) and detailed *Process Guidance Notes* (IPR 1/1,1/2, etc.) are available from HMSO, as listed in Tables 2.2a–f.

Table 2.2a IPR Notes for the fuel and power industry

IPR 1	Fuel and Power Industry
IPR 1/1	Combustion Processes: Large Boilers and Furnaces—50MW(th) and over
IPR 1/2	Combustion Processes: Gas Turbines
IPR 1/3	Combustion Processes: Compression Ignition Engines—50MW(th) and over
IPR 1/4	Combustion Processes: Waste and Recovered Oil Burners—3MW(th) and over
IPR 1/5	Combustion Processes: Combustion of Solid Fuel Manufactured from Municipal Waste in Appliances with a net rated thermal input of 3MW or more
IPR 1/6	Combustion Processes: Combustion of Fuel manufactured from, or comprised of, tyres, tyre rubber, or similar rubber waste in appliances with a net rated value of 3MW or more
IPR 1/7	Combustion Processes: Combustion of Solid Fuel manufactured from, or comprised of, Poultry Litter in appliances with a net rated thermal output of 3MW or more
IPR 1/8	Combustion Processes: Combustion/Solid Fuel which is manufactured from, or is comprised of Wood Waste or Straw in appliances with a net rated thermal input of 3MW or more
IPR 1/9	Carbonisation and associated processes: Coke Manufacture
IPR 1/10	Carbonisation and associated processes: Smokeless Fuel, Activated Carbon and Carbon Black Manufacture
IPR 1/11	Gasification Processes: Gasification of Solid and Liquid Feedstocks
IPR 1/12	Gasification Processes: Refining of Natural Gas
IPR 1/13	Gasification Processes: Refining Natural Gas at Liquefied Natural Gas (LNG) sites
IPR 1/14	Gasification Processes: The Odourising of Natural Gas or Liquefied Petroleum Gas (LPG)
IPR 1/15	Petroleum Processes: Crude Oil Refineries
IPR 1/16	Petroleum Processes: Onshore Oil Production
IPR 1/17	Combustion Processes: Reheat Furnaces and Heat Treatment 50MW(th) and over

Table 2.2b IPR Notes for the metal industry

IPR 2	Metal Industry
IPR 2/1	Iron/Steel Making Processes
IPR 2/2	Ferrous Foundry Process
IPR 2/3	Processes: Electric Arc Steel Making
IPR 2/4	Processes for the Production of Zinc and Zinc Alloys
IPR 2/5	Processes for the Production of Lead and Lead Alloys
IPR 2/6	Processes for the Production of Refractory Metals
IPR 2/7	Processes for the Production of Cadmium, Mercury and Alloys
IPR 2/8	Processes for the Production of Aluminium
IPR 2/9	Processes for the Production of Copper and Copper Alloys
IPR 2/10	Precipitation of Precious Metals/Platinum Group Metals
IPR 2/11	Extraction of Nickel/Carbonyl, Production of Cobalt/Nickel Alloys
IPR 2/12	Tin and Bismuth

Table 2.2c IPR Notes for the mineral industry

IPR 3	Mineral Industry
IPR 3/1	Cement Manufacture and Associated Processes
IPR 3/2	Lime Manufacture and Associated Processes
IPR 3/3	Processes involving Asbestos
IPR 3/4	Glass Fibres and Non-asbestos Mineral Fibres
IPR 3/5	Glass Manufacture and Production Glass Frit and Enamel Frit
IPR 3/6	Ceramic Processes

Notes Frit: calcined mixture of sand and fluxes as material for glass-making; the vitreous composition from which soft porcelain, enamel, etc., are made.

Table 2.2d IPR Notes for the chemical industry

IPR 4	The Chemical Industry
IPR 4/1	Petrochemical Processes
IPR 4/2	Processes for the Production and Use of Amines, Nitriles, Isocyanates and Pyridines
IPR 4/3	Processes for the Production or Use of Acetylene, Aldehydes etc.
IPR 4/4	Processes for the Production or Use of Organic Sulphur compounds, and Production, Use or Recovery of Carbon Disulphide
IPR 4/5	Batch Manufacture of Organic Chemicals in Multipurpose Plant
IPR 4/6	Production and Polymerisation of Organic Monomers
IPR 4/7	Processes for the Manufacture of Organo-Metallic Compounds
IPR 4/8	Pesticide Processes
IPR 4/9	Pharmaceutical Processes
IPR 4/10	Processes for the Manufacture, Use or Release of Oxides of Sulphur and the Manufacture, Recovery, Condensation or Distillation of Sulphuric Acid or Oleum
IPR 4/11	Processes for the Manufacture or Recovery of Nitric Acid and Processes involving the Manufacture or Release of Acid-forming Oxides of Nitrogen
IPR 4/12	Processes for the Sulphonation or Nitration of Organic Chemicals
IPR 4/13	Processes for the Manufacture of, or which Use or Release Halogens, mixed Halogen Compounds or Oxohalo Compounds
IPR 4/14	Processes for the Manufacture of, or which Use or Release Hydrogen Halides or any of their Acids
IPR 4/15	Processes for the Halogenation of Organic Chemicals
IPR 4/16	Processes for the Manufacture of Chemical Fertilisers or their Conversion into Granules
IPR 4/17	Bulk Storage Installations (chemicals)
IPR 4/18	Process for Manufacture of Ammonia
IPR 4/19	Production, Use, Release or Recovery of Ammonia
IPR 4/20	Production or Use of Phosphorus/Oxide/Hydride/Halide
IPR 4/21	Production or Manufacture of Hydrogen Cyanide/Sulphide
IPR 4/22	Production of Antimony/Arsenic/Beryllium/Gallium, etc.
IPR 4/23	Production, Use or Release of Cadmium Compounds, etc.
IPR 4/24	Production, Use or Release of Mercury Compounds, etc.
IPR 4/25	Production of Chromium/Magnesium/Manganese/Zinc

Notes Oleum: concentrated sulphuric acid containing excess sulphur trioxide in solution as an oily, corrosive liquid.

Table 2.2e IPR Notes for the waste disposal industry

IPR 5	Waste Disposal Industry
IPR 5/1	Merchant and In-house Chemical Waste Incineration
IPR 5/2	Clinical Waste Incineration
IPR 5/3	Municipal Waste Incineration
IPR 5/4	Animal Carcass Incineration
IPR 5/5	The Burning-out of Metal Containers
IPR 5/6	Making Solid Fuel from Waste
IPR 5/7	Cleaning and Regeneration of Activated Carbon
IPR 5/8	Recovering of Organic Solvents by Distillation
IPR 5/9	Regeneration of Ion Exchange Resins
IPR 5/10	Recovery of Oil by Distillation
IPR 5/11	Sewage Sludge Incineration

Table 2.2f IPR Notes for other industries

IPR 6	Other Industries
IPR 6/1	Application/Removal of Tributyltin/Triphenyltin
IPR 6/2	Tar and Bitumen Processes
IPR 6/3	Timber Preservation Processes
IPR 6/4	Di-isocyanate Manufacture
IPR 6/5	Toluene Di-isocyanate Use/Flame, Polyurethane
IPR 6/6	Textile Treatment Processes
IPR 6/7	Processing of Animal Hide and Skin
IPR 6/8	Making of Paper Pulp by Chemical Methods
IPR 6/9	Papermaking and Related Processes

Each of these *Process Guidance Notes* is organised in a similar format, the key elements of which are as follows:

Main text

- Process Description
- Releases into Air
- Releases into Water
- Releases into Land

Annexes

- Processes
- Prescribed Substances
- Techniques (Abatement Techniques—BATNEEC)
  - — For release into Air
  - — For release into Water
  - — For release into Land

- Compliance Monitoring Programme
- Additional Requirements.

Examples of BATNEEC

BATNEEC (IPR 1/1): Large boilers and furnaces (50MW) and over

Releases into air

Table 2.3 BATNEEC: large boilers and furnaces (50MW) and over—releases into air

Substance	Process	BATNEEC
Halogens and their compounds	Combustion of fuels (esp. coal)	Wet washing of combustion gases
Metals, metalloids and their compounds	Combustion of coal, residual fuel oils or bitumen emulsions	Pulverised coal firing, electrical precipitators, fabric filters
Nitrogen oxides	Firing of oil and solid fuels	Burner/combustion chamber design, gasification techniques, selective non-catalytic reduction, combined SO_2/NOx processes, advanced combustion technologies
Organic compounds	Incomplete combustion	Proper combustion control, combustion zone design
Oxides of carbon	Combustion of fuels	Absorption/desorption techniques (not currently a practicable option); CO should be reduced to CO_2 by good combustion practice
Particulate matter: ash particles, condensate droplets	Coal firing, firing heavy fuel, oil or oil/water emulsions	Pulverised coal firing, fabric filters
Sulphur dioxide	Combustion of fuel containing sulphur	Flue gas desulphurisation equipment

Instrumentation and monitoring

In general, continuous monitoring and recording equipment for the final discharge to atmosphere of combustion gases should be installed on all new plant. Substances to be measured (2) are:

- particulate matter
- sulphur dioxide
- nitrogen oxides

Additional requirements

These cover cooling towers, materials handling, commissioning, general operations, chimneys and vents, and include the following requirements:

- Cooling towers
  — Droplet release from cooling towers should be monitored and prevented
  — Cooling towers should not be of the evaporative type

Materials handling

- Emission of acid soot should be prevented
- The consistencies of oil fuels should be monitored
- Combustion passages must be airtight
- Start-up emissions should be identified
- The release of prescribed substances, smells or harmful circumstances arising from liquid fuels should be prevented
- The emission of dust to air from solid fuels should be prevented
- Stocks of dusty materials should be stored under cover and operations should be carried out so as to minimise emissions to air
- Suitable dust-arresting systems should be incorporated so that an emission level of $50mg/m^3$ is not exceeded
- Storage silos for dusty materials should be vented to air so that an emission level of $50mg/m^3$ is not exceeded

Commissioning trials must be conducted and reported to demonstrate plant performance capability.

General operations guidance details requirements for qualified personnel, education, maintenance, supervision, the handling of breakdowns and emergencies, and loose oil capture:

- Chimneys and vents must ensure that emissions are free from odours and harmless at ground level
- Chimneys should be of the correct height to disperse emissions, particularly nitrogen oxides, hydrogen chloride, sulphur dioxide, and other pollutants
- Efflux velocities should be greater than 15 m/s at full load operation and flue gases, where a wet method of arrestment is used, should be reheated to exceed 80°C so as to minimise the generation of a steamy plume

- Flues and ductwork should be insulated to minimise the cooling of gases and to prevent condensation on internal surfaces
- To maximise thermal buoyancy, the minimum practicable number of chimneys should be used

Releases into water

Table 2.4 BATNEEC: large boilers and furnaces (50MW) and over—releases into water

Substance	Process	BATNEEC
Ash quenching water	Combustion processes	May need chemical treatment before discharge Two-stage hydrogen-peroxide precipitation at different pH values Sulphide precipitation to remove cadmium and mercury
Blow-down water	Combustion processes	Normally has no environmental significance
Cleaning liquids	Combustion processes	All these should be neutralised or treated to produce an acceptable waste before discharge or disposal
Cooling tower water purge	Combustion processes	Various substances are available to prevent the formation of slime and other organisms
De-ionisation effluent	Combustion processes	This should be neutralised before discharge
Liquid effluent	Post-combustion power plant	May need chemical treatment before discharge Two-stage hydrogen peroxide precipitation at different pH values
Surface water	Storage of solid fuels	May need treatment to remove oil, soot, dissolved hydrocarbons

Releases into land

Table 2.5 BATNEEC: large boilers and furnaces (50MW) and over—releases into land

Substance	Process	BATNEEC
Heavy metals	Oil or oil/emulsion fired boilers	Requires special handling
Solid matter: fly ash, bottom ash, slag dust	Combustion processes	(May be saleable) Collect in dry arrestment plant Quenching into tanks with scraper conveyers
Sulphur or gypsum	Flue gas desulphurisation	(May be saleable)
Waste catalyst	Selective catalytic reduction processes	(May be saleable)

BATNEEC (IPR 2): metal industry sector

- The cleanest techniques should be used
- Waste should be minimised by the efficient recycling of materials
- Dirty water should be cleaned and reused
- Contaminated air should be used for process combustion air
- Solid wastes should be reprocessed

Releases into air, water and land

Table 2.6 BATNEEC: metal industry sector—releases into air, water and land

Substance	Process	BATNEEC
Particulate matter	Metals processing	Contain sources Collect dust or fume and duct to an efficient arrestment system (bag filter, electrostatic precipitator or multicyclone separators)
Sulphur dioxide	Use of fuel for metals processing	Convert to sulphuric acid Use lime or caustic scrubbing
Oxides of carbon	Use of fuel for metals processing	Burn CO to CO_2 Increase thermal efficiencies to reduce CO_2 production

Table 2.6 *(continued)*

Substance	Process	BATNEEC
Hydrocarbons and smoke	Use of fuel for metals processing	Use clean raw materials Adopt careful combustion conditions Incinerate at a temperature exceeding 800°C in oxidising conditions
Fluorides	Metals processing	Water scrubbing Remove hydrogen fluoride by contact with lime or limestone dust which may be insufflated into the gas flow upstream of a bag filter
Arsine and stibine	Metals processing	May be avoided by not using reducing or damp conditions
Ammonia	Aluminium processing	Aluminium drosses should not be allowed to become wet before processing for the removal of aluminium as they may emit ammonia
Chlorine and its compounds	Metals processing	Scrub out with water or caustic solutions
Phosphorus and its compounds	Phosphor copper processing	Avoid alloys above 15% phosphorous Scrub using a water-irrigated candle mist arrester or a high pressure drop venturi scrubber
Mercury in water	Metals processing	React with sodium sulphide followed by flocculation-enhanced settlement and a filtration or polishing process Recycle the sulphide for recovery of the metal
Cadmium in water	Metals processing	Precipitation by the use of lime or caustic followed by flocculation-enhanced settlement and a filtration or polishing process Recycle the solids collected

Instrumentation and monitoring

Continuous sampling and monitoring of the following is required:

- particulate matter in chimneys and ducts
- instantaneous and average effluent flows
- effluent compositions

General BATNEEC

The specific IPR booklets should be consulted for the full and comprehensive details of the pollution abatement techniques. There follows a general list of pollutant emissions to air, water and land, some common BATNEEC clean-up techniques and measurements required to verify compliance.

Releases into air—general principles

The foremost requirement to minimise releases into air is containment. Particulate matter should be collected and ducted to an efficient arrestment system, such as bag or ceramic filters, electrostatic precipitators or multicyclone separators. Damping may help the containment of dust but makes ducting difficult and may result in clogged filters. Acid gases may be removed by wet scrubbing using neutralising alkaline slurries, which may be evaporated to dryness and ducted to downstream particulate matter abatement plant. Pollutants may be adsorbed on to a solid matrix, absorbed into a liquid or incinerated out.

Table 2.7 General BATNEEC—releases into air

Substance	BATNEEC
Acid gases	Wet scrubbing (e.g. venturi or packed or tray tower) Lime or sodium carbonate slurries (with pH control) Evaporated to dryness Dry scrubbing: powdered lime injection Downstream particulate matter abatement plant (fabric or ceramic filters) if necessary
Asbestos	Containment and dampening
Cadmium	Sulphide precipitation, see particulate matter
Carbon dioxide	See oxides of carbon
Carbon monoxide	See oxides of carbon
Condensate droplets	See particulate matter Maintain temperatures above dew-point
Dioxins	Carbon injection in a dry scrubber (adsorption by carbon)
Fumes	Fume arrestment plant usually scrubs the gas using a water-irrigated candle mist arrester or a high pressure drop venturi scrubber

Table 2.7 *(continued)*

Substance	BATNEEC
Gaseous organic pollutants	Condensation Adsorption (on to a solid matrix) (e.g. activated carbon, bauxite, magnesia) Absorption (into a liquid) Combustion (flaring or controlled incineration)
Halogens and their compounds	Wet washing of combustion gases
Hydrocarbons	Use clean raw materials and careful combustion conditions
Hydrogen chloride	Insufflation of ground limestone
Hydrogen fluoride	Remove hydrogen fluoride by contact with lime or limestone dust which may be insufflated into the gas flow upstream of a bag filter
Lead	See particulate matter
Metals, metalloids and their compounds	Pulverised coal firing Electrical precipitators Fabric filters
Mercury	Carbon injection in a dry scrubber (adsorption by carbon) Sulphide precipitation
Nitrogen oxides	Burner/combustion chamber design Combustion control Gasification techniques Selective catalytic reduction (SCR) (using ammonia gas) Combined SO_2/NOx processes Advanced combustion technologies Steam injection Water injection Urea injection (check ammonia produced) Control the combustion at a lower temperature by staged combustion techniques
Odours	Regular cleaning and disinfection Limit waste accumulation Air extraction Wet scrubbing with alkaline sodium hypochlorite Biological scrubbing: the suspension of micro-organisms in an aqueous solution or as a growing culture on a solid carrier Biofiltration: involves absorption, condensation and biological oxidation Use of afterburners
Organic compounds	Proper combustion control Combustion zone design
Oxides of carbon	Absorption/desorption techniques (not currently a practicable option) CO should be reduced to CO_2 by good combustion practice Increase thermal efficiencies to reduce CO_2 production
Oxides of nitrogen	See nitrogen oxides

continued overleaf

Table 2.7 *(continued)*

Substance	BATNEEC
Particulate matter, ash particles	Contain sources Pulverised coal firing Collect dust or fume and duct to an efficient arrestment system: bag filter Ceramic filters (for high temperature corrosive gases) Electrostatic precipitator Multicyclone separators Bag 'blinding' or filter fires may be prevented by the insufflation of ground limestone Wet scrubber water sprays Gravel bed filter
Smoke	Adopt careful combustion conditions Incinerate at a temperature exceeding 800°C in oxidising conditions Use clean raw materials Filtration (may cause filter bag blinding or jet blockages) Wet scrubbing (may cause filter bag blinding or jet blockages)
Sulphur dioxide	Flue gas desulphurisation equipment Use lime or caustic scrubbing Convert to sulphuric acid Use gas or low sulphur content oil
Wet, hot or aggressive gases	Venturi scrubbers Impeller disintegrators

Notes Venturi scrubber: in this device, the gas is forced through a venturi throat in which the gas is mixed with high-pressure liquid sprays. Insufflation: blowing in of dust. 'Blinding': filling up the pores.

Instrumentation and compliance monitoring requirements

- Ambient air temperature
- Amines
- Ammonia
- Carbon dioxide—flue gas analysis equipment
- Carbon monoxide—flue gas analysis equipment monitors for carbon monoxide are usually based on radiation absorption in the infra-red band
- Combustion chamber gas temperatures (incinerators)
- Dioxins
- Fugitive releases (e.g. smells, leaks)
- Gas flows
- Heavy metals
- Hydrogen chloride—process photometer (infra-red absorption); contact sample gas with an absorbing solution with subsequent measurement of the chloride ion concentration by means of an ion-sensitive electrode

- Hydrogen fluoride—contact sample gas with an absorbing solution with subsequent measurement of the fluoride ion concentration by means of an ion-sensitive electrode
- Hydrogen sulphide
- Nitrogen oxides—the value to be monitored is the nitrogen dioxide equivalent of nitric oxide and nitrogen dioxide present
- Oxygen content—flue gas analysis equipment
- Particulate matter—optical density, beta ray, particulate impingement, light scattering, continuous sampling devices
- Prescribed substances
- Pressure
- Smells
- Sulphur dioxide—monitors for sulphur dioxide are usually based on radiation absorption in the infra-red band
- Temperature
- Volatile organic compounds—flame ionisation detectors introduce a sample of the gas into a hydrogen flame in an electric field; ionisation of the flame, and hence the current flow, is proportional to the number of carbon atoms present with carbon-hydrogen bonds
- Volatile organic sulphur compounds
- Water content
- Substances required by LAAPC

Releases into water—general principles

Suspended solids may be removed by settlement and/or filtration. Effluents may need chemical treatment for neutralisation. Heavy metals may by removed by chemical precipitation or flocculation followed by settlement and/or filtration. The solids collected may be recycled. Highly contaminated effluent may be combusted.

Table 2.8 General BATNEEC—releases into water

Substance	BATNEEC
Ammonia	Aluminium drosses should not be allowed to become wet before processing for the removal of aluminium as they may emit ammonia
Arsine and stibine	May be avoided by not using reducing or damp conditions
Asbestos	Filtration
Ash quenching water	May need chemical treatment before discharge Two-stage hydrogen peroxide precipitation at different pH values Sulphide precipitation to remove cadmium and mercury

continued overleaf

Table 2.8 *(continued)*

Substance	BATNEEC
Blow-down water	Normally has no environmental significance, although may contain phosphates, alkalis, hydrazine, ammonia etc., used for pH control, de-aeration, etc.
Cadmium in water	Precipitation by the use of lime or caustic solution followed by flocculation-enhanced settlement and a filtration or polishing process Recycle the solids collected
Chlorine and its compounds	Scrub out with water or caustic solutions
Cleaning liquids	All these should be neutralised or treated to produce an acceptable waste before discharge or disposal
Cooling tower water purge	Various substances are available to prevent the formation of slime and other organisms
De-ionisation effluent	This should be neutralised (pH, soluble sulphates) before discharge
Fluorides	Water scrubbing
Lead	Precipitation and filtering
Liquid effluent	May need chemical treatment before discharge Two-stage hydrogen peroxide precipitation at different pH values Sulphide precipitation to remove cadmium and mercury
Liquid effluent from gas treatment plant	Filtration Chemical treatment Two-stage hydroxide precipitation at different pH values Sulphide precipitation to remove cadmium and mercury
Metals	Chemical precipitation processes followed by the removal of suspended particulate matter Acid neutralisation and precipitation with calcium hydroxide can remove up to 90% of most heavy metals but probably less than 70% of cadmium and nickel
Mercury in water	React with sodium sulphide followed by flocculation-enhanced settlement and a filtration or polishing process Recycle the sulphide for recovery of the metal Specialist precipitation agents can remove up to 99% of the metal
Odoriferous compounds	See sulphides
Phosphorus and its compounds	Avoid alloys above 15% phosphorus Scrub using a water-irrigated candle mist arrester or a high pressure drop venturi scrubber
River water	See Chapter 6
Sulphides	Containment and incineration Scrubbing with reactive liquor
Surface water	May need treatment in separation/interceptor systems to remove dissolved hydrocarbons Effluent suspended solids interceptor system Treatment to remove dissolved pollutants

Table 2.8 *(continued)*

Substance	BATNEEC
Suspended particulate matter	Settlement, filtration, centrifuges (a preliminary chemical precipitation process may be required)
Tap water	See Chapter 6
Waste water (heavily contaminated)	Oil/Water separation Neutralisation Heavy metal removal Solids removal Incineration

Instrumentation and compliance monitoring requirements

- BOD (biochemical oxygen demand)
- COD (chemical oxygen demand)
- Effluent sampling—continuous and flow proportional
- Effluent compositions
- Fugitive releases (e.g. leaks)
- Instantaneous and average effluent flows
- pH levels
- Prescribed substances
- Temperature
- TOC (total organic carbon)
- Toxicity
- Turbidity (suspended solids)
- Substances required by the NRA

Releases into land—general principles

Solid waste should be rendered harmless, packaged and clearly labelled and passed over to a registered waste disposal contractor.

Table 2.9 General BATNEEC—releases into land

Substance	BATNEEC
Asbestos	Render harmless: packaging, labelling and notification
Dust	Containment, bag filter
Heavy metals	Require special handling
Solid matter	(May be saleable)
Fly Ash	Collect in dry arrestment plant
Bottom ash	Quenching into tanks with scraper conveyers

continued overleaf

Table 2.9 *(continued)*

Substance	BATNEEC
Slag	Containment (may be saleable)
Sludges	Incineration Stabilisation and solidification prior to disposal Direct disposal to a licensed site
Sulphur or gypsum	(May be saleable)
Waste catalyst	(May be salelable)

Instrumentation and compliance monitoring requirements

- Flora
- Fugitive releases (e.g. spills)
- Prescribed substances
- Quantities of waste
- Soil

REFERENCES

(1) R. Welford and A. Gouldson, *Environmental Management and Business Strategy*, Pitman Publishing, London, 1993.
(2) *Croners Environmental Management*, Croner Publications, London, 1991.
(3) Department of the Environment and the Welsh Office, *Integrated Pollution Control: A Practical Guide*, HMSO, London, 1993.
(4) Her Majesty's Inspectorate of Pollution, *Chief Inspector's Guidance to Inspectors, Environmental Protection Act 1990*, IPR Process Guidance Notes, HMSO, London, 1991

APPENDIX 2 PRESCRIBED PROCESSES AND SUBSTANCES

The **Environmental Protection (Prescribed Processes and Substances) Regulations 1991** lists Part A and Part B processes under the following headings, which are intended as an aide mémoire only. The original should be consulted in conjunction with the EPA for full details.

A2.1 PART A PROCESSES

Part A processes are the most significant polluting processes in the UK and all of their emissions are controlled under the EPA by Her Majesty's Inspectorate of Pollution, HMIP, in England and Wales. These processes are obliged to apply the objectives of integrated pollution control (IPC), whereby the total emissions to air, water and land are considered under a single application for authorisation to emit. The company is required to minimise its overall environmental impact by ensuring that, and convincing HMIP that, the best practicable environmental options (BPEOs) are achieved.

The objective is that the best available techniques not entailing excessive costs (BATNEEC) must be used to prevent or minimise pollutant releases to the environment. If authorisation to emit is to be granted, the emitter must satisfy the relevant authority that the process for which they are responsible applies the BATNEEC in the control of emissions.

The production of fuel and power and associated processes

Table 2.10 Part A prescribed processes in the production of power and associated processes

Processes / Substances	Natural gas, gas	LPG	Coal	Lignite	Oil	OCM	Waste
Gasification			×	×	×	×	
Producing[0]	×						
Reforming	×		×	×	×	×	
Refining	×	×	×	×	×	×	
Odorising	×	×					
Purifying	×	×	×	×	×	×	
Pyrolysis[1]			×	×	×	×	
Carbonisation[1]			×	×	×	×	
Distillation[1]			×	×	×	×	
Liquefaction[1]			×	×	×	×	
Partial oxidation[1]			×	×		×	
Heat treatment[1]			×	×	×	×	
Conversion	×	×	×	×	×	×	
Combustion (50MW or more)		×	×	×	×		
Combustion (3MW or more)					×[2]	×	×
Loading					×[3]		
Unloading					×[3]		
Handling					×[3]		
Storage					×[3]		
Treatment					×[3]		

Notes

LPG—liquid petroleum gas

OCM—other carbonaceous material, such as charcoal, coke, peat and rubber

(0) Other than from sewage or the biological degradation of waste.

(1) Other than gasification or the making of charcoal.

(2) Waste oil.

(3) Crude oil, stabilised crude petroleum, crude shale oil, and any associated gas or condensate.

Metal production and processing

Iron and steel

Table 2.11 Part A prescribed processes in metal production and processing—iron and steel

Substances / Processes	Iron ore	Burnt pyrites	Iron	Steel	Ferro-alloy
Loading	×[4]	×			
Unloading	×[4]	×			
Handling	×[4]	×			
Storing	×[4]				
Crushing	×				
Grading	×				
Grinding	×				
Screening	×				
Washing	×				
Drying	×				
Blending	×				
Mixing	×				
Pelletising	×				
Calcining	×				
Sintering	×				
Making			×	×	×
Melting			×	×	×
Refining			×	×	×
Desulphurisation			×	×	×
Heating[5]			×	×	×
Foundrying			×	×	×
Handling slag			×	×	×

Notes

(4) Except in the course of mining operations.

(5) To remove grease, oil, or contaminant.

Non-ferrous metals

Table 2.12 Part A prescribed processes in metal production and processing—non-ferrous metals

Processes / Substances	A	Zinc[6]	Tin[6]	B	C	D	E
Extracting	×						
Recovering	×			×	×		
Mining[7]		×	×				
Refining[8]	×						
Making[9]	×						
Melting[9]	×			×	×		
Producing				×	×		
Manufacturing						×	
Repairing						×	
Heating[10]	×						
Foundrying	×	×	×	×	×	×	×
Smelting							×
Calcining							×

Notes

A. Any non-ferrous metal or alloy if the process may result in the release into the air of particulate matter or any metal, metalloid (or compounds of these), or in the release into water of a substance described in SCHEDULE 5 of the Regulations, listed later.

B. When the process produces a molten alloy which contains a proportion which exceeds the relevant percentage of the following substances, and the process may result in the release into the air of particulate matter or smoke which contains any of these elements:

antimony	1%
arsenic	1%
beryllium	0.1%
chromium	2%
lead when alloyed with copper	23%
lead when alloyed with any metal other than copper	2%
magnesium	10%
manganese when alloyed with copper	15%
manganese when alloyed with any metal other than copper	4%
phosphorus	1%
platinum	1%
selenium	0.5%

C. Cadmium or mercury or any alloy containing more than 0.05% by weight of either of these metals or both of these metals in aggregate.

D. Beryllium or selenium or an alloy of one or both of these metals if the process may occasion the release into the air of any substance described in SCHEDULE 4 listed later.

E. Sulphides, sulphide ores, including regulus or mattes.

(6) Where the process may result in the release into water of cadmium or compounds of cadmium.

(7) Where the process may result in the release into water of cadmium or compounds of cadmium.

(8) Except where the process falls within a Part B process.

(9) When the processing vessel has a capacity of 5 tonnes or more.

(10) To remove grease, oil, or contaminant.

Mineral industries

Table 2.13 Part A prescribed processes in mineral industries

Processes / Substances	Cement/ clinker	Ready-mixed concrete	Concrete blocks and products	F	Lime	Asbestos and products	Glass fibre[11]	G	H
Manufacturing	×		×			×	×		
Grinding	×								
Blending	×								
Storing (bulk silos)	×								
Removing from above	×								
Bagging in bulk	×								
Batching		×							
Heating[12]				×					
Slaking[13]					×				
Stripping						×			
Firing								×	
Vapour glazing									×

Notes

F. Calcium carbonate or calcium magnesium carbonate.

G. Heavy clay goods or refractory goods in a kiln where a reducing atmosphere is used for purposes other than coloration.

H. Earthenware or clay with salts.

(11) Or any fibre from any mineral other than asbestos and glass frit or enamel frit.

(12) For the purpose of making lime.

(13) For the purpose of making calcium hydroxide or calcium magnesium hydroxide.

The chemical industry

Table 2.14 Part A prescribed processes in chemical industries

Substances / Processes	Olefins	I	J	K	L	M	N	O	P	Q	R
Manufacture	×	×	×	×	×	×	×	×	×	×	×
Use of	×	×	×	×	×	×	×	×	×	×	×
Recovery		×		×	×		×	×			
Polymerisation	×	×									
Release of			×	×				×[14]	×		
Purification							×				
Concentration								×			
Distillation								×			
Conversion of										×	

Notes

I. Styrene or vinyl chloride.

J. Acetylene, aldehyde, amine, isocyanate, nitrile, organic acid or its anhydride, any organic sulphur compound, or any phenol; if the process may result in the release of any of these substances into the air.

K. Carbon disulphide.

L. Pyridine, methyl pyridine, or di-methyl pyridine.

M. Organo-metallic compounds.

N. Acrylate.

O. Acids: sulphuric acid or oleum, oxides of sulphur (except combustion or incineration processes other than the burning of sulphur), nitric acid, acid-forming oxides of nitrogen, or phosphoric acid.

P. Fluorine, chlorine, bromine, iodine, or any compound comprising two or more of these halogens (or one or more of these with oxygen); hydrogen fluoride, hydrogen chloride, hydrogen bromide, hydrogen iodide or any of their acids, hydrogen cyanide, hydrogen sulphide (other than in fumigation), any of the following and their compounds:

antimony	arsenic	beryllium	gallium	indium	lead
palladium	platinum	selenium	tellurium	thallium	

where the process may result in the release in to air of any of these elements or compounds or the release into water of any substance described in SCHEDULE 5 (listed later); cadmium or mercury; any compound of chromium, magnesium, manganese, nickel, zinc; any metal carboyl, ammonia, phosphorus and its oxides, hydrides and halides.

Q. Chemical fertilisers and pesticides if the process may result in the release into water of any substance described in SCHEDULE 5 (listed later) or if 500 tonnes/annum, or more, of special waste might be produced by the operation.

R. Medicinal products if the process may result in the release into water of any substance described in SCHEDULE 5 below or if 1000 tonnes/annum, or more, of special waste might be produced by the operation.

(14) Except in the combustion of carbonaceous material.

The storage of the following substances become a Part A process if the storage capacity exceeds the values indicated below:

Substance	**Tonnes**
Any one or more acrylates	20
Acrylonitrile	20
Anhydrous ammonia	100
Anhydrous hydrogen fluoride	1
Toluene di-isocyanate	20
Vinyl chloride monomer	20

Waste disposal and recycling

Incineration of

- Waste chemicals or waste plastic arising from manufacture.
- Bromine, cadmium, chlorine, fluorine, iodine, lead, mercury, nitrogen, phosphorus, sulphur or zinc.
- Any other waste, including animal remains, otherwise than by a process related to a Part B process, on premises where there is plant designed to incinerate such waste at a rate of 1 tonne or more per hour.
- The cleaning for reuse of metal containers used for the transport or storage by burning out their residual content.

Recovery processes

Recovery by distillation of any oil or any organic solvent or the cleaning or regeneration of carbon, charcoal or ion exchange by removing substances described in SCHEDULES 4, 5 and 6 listed later.

The production of fuel from waste

Any such process which involves the use of heat other than making charcoal.

Other industries

- Paper and pulp manufacturing processes (if the site capacity is greater than 25 000 tonnes of paper pulp or related processes, or which may result in the release of any SCHEDULE 5 substances listed later).
- Processes involving the manufacture or release of di-isocyanates, including the hot-wire cutting, thermal debonding or flame bonding of polyurethane foams or polyurethane elastomers.
- Tar and bitumen processes (e.g. distillation or heating 5 tonnes or more in a 12-month period).
- Processes involving uranium.
- Coating processes and printing (e.g. involving tributylin or triphenyltin compounds).
- The treatment of textiles where SCHEDULE 5 substances may be released, the application of printing ink or paint where greater than 1000 tonnes of special waste may be produced in any 12-month period.
- The manufacture of dyestuffs, printing ink and coating materials, where hexachlorobenzene is used, or where greater than 1000 tonnes of special waste might be produced in a 12-month period.

- Timber processes, such as curing or chemical treatment if any SCHEDULE 5 substance is used or the use of wood preservatives where greater than 500 tonnes of special waste may be produced in any 12-month period.
- Processes involving rubber such as mixing, milling or blending, if carbon black is used.
- The treatment and processing of animal or vegetable matter where SCHEDULE 5 substances may be released.

A2.2 PART B PROCESSES

Part B processes come under the control of Local Authorities, under LAAPC (Local Authority air pollution control). Although the latter are only obliged to regulate emissions to air, the processes involved are still bound to apply the principles of BATNEEC.

The production of fuel and power and associated processes

Table 2.15 Part B prescribed processes in the production of power and associated processes

Processes \ Substances	Any fuel	Waste oil	Fuel from waste
Burning	20–50 MW	less than 3MW	less than 3MW

Metal production and processing

Iron and steel

Table 2.16 Part B prescribed processes in metal production and processing—iron and steel

Processes \ Substances	Iron	Steel	Ferro-alloy
Making	×	×	×
Melting	×	×	×
Refining	×	×	×
Desulphurisation	×	×	×

Non-ferrous metals

Table 2.17 Part B prescribed processes in metal production and processing—non-ferrous metals

Substances / Processes	Non-ferrous metal	Non-ferrous alloy	S
Making	less than 5 tonnes	less than 5 tonnes	
Melting	less than 5 tonnes	less than 5 tonnes	×
Extraction			×
Recovery			×

Notes
S. Copper, aluminium or zinc from mixed scrap using heat.

Mineral industries

Table 2.18 Part B prescribed processes in mineral industries

Substances / Processes	Cement[15]	Cement clinker[15]	Lime	Any mineral material[16]	Coal or coke	Bricks, tiles or concrete	Road-stone	Glass	Heavy clay goods
Storing	×	×							
Loading	×	×			×[18]				
Unloading	×	×			×				
Blending	×[17]								
Using	×								
Slaking			×						
Crushing				×	×	×			
Grinding				×	×	×			
Grading				×	×				
Screening				×	×	×			
Heating				×					
Breaking up					×				
Mixing					×				
Coating[19]							×		
Manufacture								×[20]	
Firing									×

Notes
(15) In bulk prior to further transportation in bulk.
(16) Except where the process is unlikely to release into the air any particulate matter.
(17) In bulk other than at a construction site.
(18) Except on retail sale.
(19) With tar or bitumen.
(20) Where the person concerned has the capacity to make 5000 tonnes or more in any 12-month period, or where lead or lead compounds are used, or polishing or etching if hydrofluoric acid is used, or hydrogen fluoride may be released into the air.

Waste disposal and recycling

The destruction by burning of any waste, including animal remains and the cremation of human remains.

Other industries

- Any process which is likely to involve the use of 5 tonnes or more of di-isocyanates or partly polymerised di-isocyanates in any 12-month period.
- Coating processes and printing.
- Processes which are likely to produce 1000 tonnes or more of special waste within any 12-month period where the process does not come under Part A, or when the process may result in the release into the air of particulate matter or any volatile organic compound, or when the process may involve 20 tonnes or more of printing ink, paint or other coating material, or 20 tonnes or more of any metal coating, or 5 tonnes of organic solvents in any 12-month period.
- Any process for the respraying of road vehicles if the process may result in the release into the air of particulate matter or any volatile organic material and the carrying on of the process at the location may involve the use of 2 tonnes or more of organic solvents in any 12-month period.
- The manufacture of dyestuffs, printing ink and coating materials.
- Any process for the manufacture or formulation of printing ink or any other coating material containing, or involving the use of 100 tonnes or more of any organic solvent in a 12-month period.
- Any process for the manufacture of any powder for use as a coating where there is capacity to produce 200 tonnes or more of such powder in any 12-month period.
- Timber processes.
- Processes where greater than 10 000 m^3 of wood is sawed but not otherwise processed at the works, or in any other case, is likely to exceed 500 m^3 in any 12-month period.
- Processes involving rubber.
- The mixing, milling or blending of natural rubber or synthetic elastomers if carbon black is used.
- The treatment and processing of animal or vegetable matter.
- Any Part A process where the process may not release into water any SCHEDULE 5 substance, but may release into the air a SCHEDULE 4 substance or any offensive smell noticeable outside the premises.
- Breeding of maggots where more than 5 kg of animal or vegetable matter (or in aggregate) may be introduced into the process in any week.

Prescribed substances

Release into the air (SCHEDULE 4)

- Oxides of sulphur and any sulphur compounds
- Oxides of nitrogen and any nitrogen compounds
- Oxides of carbon
- Organic compounds and partial oxidation products
- Metals, metalloids and their compounds
- Asbestos (suspended particulate matter and fibres), glass fibres and mineral fibres
- Halogens and their compounds
- Phosphorus and its compounds
- Particulate matter

Release into water (SCHEDULE 5)

- Mercury and its compounds
- Cadmium and its compounds
- All isomers of hexachlorocyclohexane
- All isomers of DDT
- Hexachlorobenzene
- Hexachlorobutadiane
- Aldrin
- Dieldrin
- Endrin
- Polychlorinated biphenyls
- Dichlorvos
- 1,2-Dichloroethane
- All isomers of trichlorobenzene
- Atrazine
- Simazine
- Tributyltin compounds
- Triphenyltin compounds
- Trifluralin
- Fenitrothion
- Azinphos-methyl
- Melathion
- Endosulfan

Release into land (SCHEDULE 6)

- Organic solvents
- Azides

- Halogens and their covalent compounds
- Metal carbonyls
- Organo-metallic compounds
- Oxidising agents
- Polychlorinated dibenzofuran and any congener thereof
- Polychlorinated dibenzo-p-dioxin and any congener thereof
- Polyhalogenated biphenyls, terphenyls and naphthalenes
- Phosphorus
- Pesticides
- Alkali metals and their oxides and alkaline earth metals and their oxides

Do you come under the provisions of the EPA?

Examine SCHEDULES 4, 5 and 6. Do you have any of these substances on site? Does your site release any of these substances to the environmental media: air, water or land?

Which industry are you in?

- The production of fuel and power and associated processes
- Metal production and processing
- Mineral industries
- The chemical industry
- Waste disposal and recycling
- Other industries:
  - paper and pulp manufacturing processes
  - processes involving the manufacture or release of di-isocyanates
  - tar and bitumen processes
  - processes involving uranium
  - coating processes and printing
  - the treatment of textiles
  - the manufacture of dyestuffs, printing ink and coating materials
  - timber processes
  - processes involving rubber
  - the treatment and processing of animal or vegetable matter

Refer to the relevant section in this Appendix to obtain an indication of your position with regard to IPC registration and authorisation. Take particular note of the emission thresholds stated.

Check the Environmental Protection Act, 1990 and the Environmental Protection (Prescribed Processes and Substances) Regulations, 1991 (both available from HMSO, Crown Copyright).

If you decide that you need to apply for authorisation to emit under the EPA IPC, then obtain the Environmental Protection (Applications, Appeals and Registers) Regulations, 1991 from HMSO and follow the procedures outlined there.

3

Environmental standards: going green?

This chapter describes the European Union's Eco-management and Audit Scheme which allows voluntary participation by companies, so that they may assess their environmental impacts, set policy and plan for improvements. The concepts of environmental policies, environmental objectives, environmental programmes, environmental management systems, environmental review, environmental auditing, environmental statements and environmental information disclosure are introduced. The essential components are listed of the British Standards Institution's specification for Environmental Management Systems, BS 7750:1992, which is applicable to any organisation which wishes to assure itself of compliance with a stated environmental policy, and demonstrate such compliance to others. The voluntary EC Environment Council Ecolabelling Regulation 1992, intended to promote the design, production, marketing and use of products which have a reduced environmental impact during their entire life cycle and to provide consumers with better information on the environmental impact of products, is examined and its implications discussed. The concepts of product environmental audits, cradle-to-grave environmental impacts of products, life cycle analysis and life cycle assessment are introduced.

3.1 THE EUROPEAN UNION'S ECO-MANAGEMENT AND AUDIT SCHEME (EMAS)

In 1991, the European Commission approved a proposal for a Council Regulation to establish a European Community Eco-management and Audit Scheme (Figure 3.1). The Council Regulation (1), published in July 1993, allowed voluntary participation by companies in the industrial sector in such a scheme, which provides a framework for companies to assess their environmental impacts, set policy and plan for improvements. Whilst participation is initially voluntary, it is expected that it will become compulsory for larger companies in time.

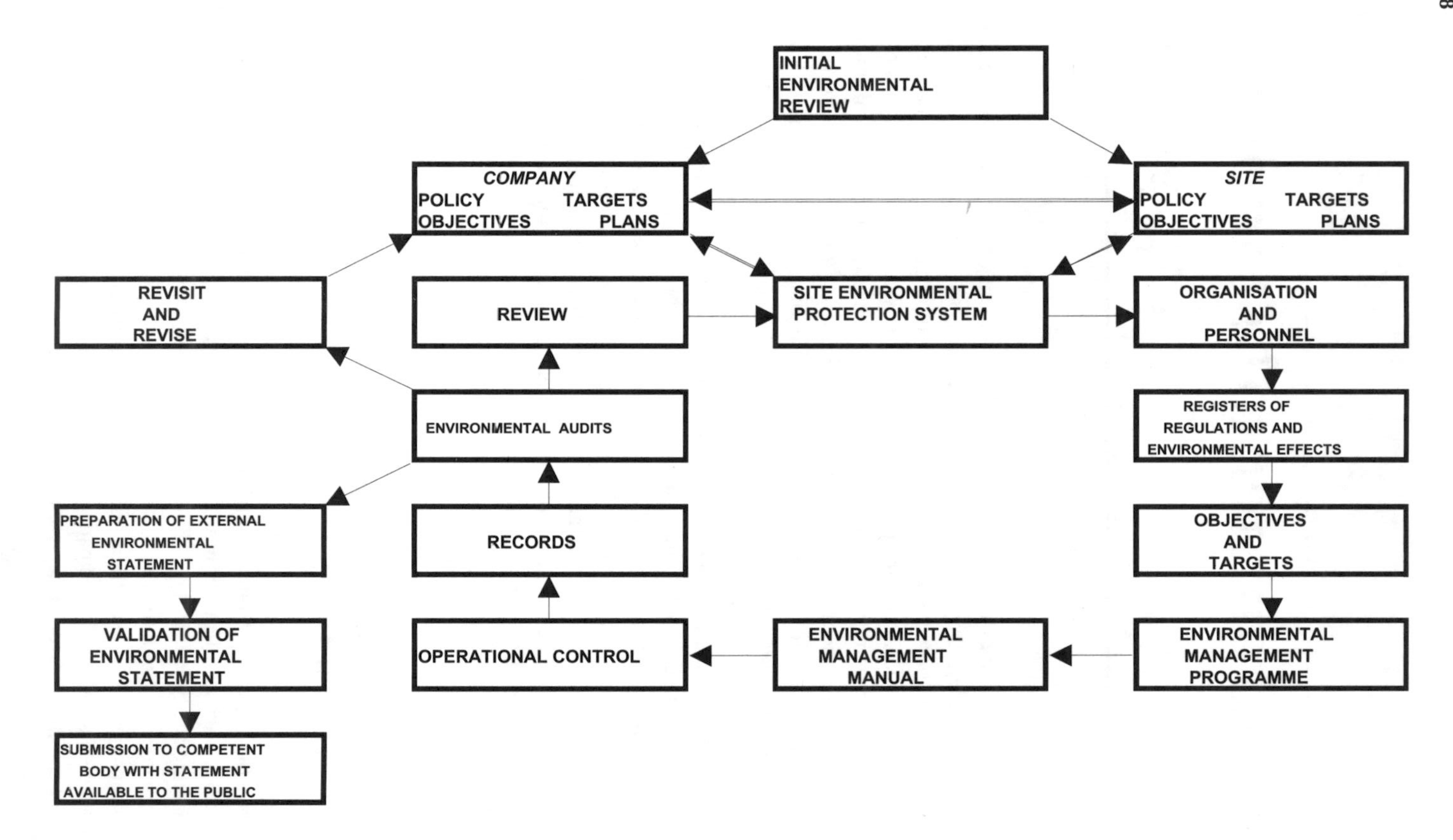

Figure 3.1 EU Eco-management and Audit Scheme

The scheme called for companies to establish and implement:

- *environmental policies* providing for compliance with all relevant regulatory requirements and making commitments aimed at the reasonable continuous improvement of environmental performance;
- *environmental objectives*;
- effective *environmental management systems*, which include ensuring awareness and training of workers in the establishment and implementation of these systems;
- *environmental auditing* procedures to help management assess compliance with the environmental management system and the effectiveness of the latter in fulfilling the company's environmental policy;
- systems of *environmental information disclosure* to the public.

It would be necessary to have the independent and neutral accreditation and supervision of environmental verifiers to ensure the credibility of the scheme.

The concept of an *environmental review* was established, as meaning *an initial comprehensive analysis of the environmental issues, impact and performance related to activities at a site.*

The *environmental audit* was defined as *a management tool comprising a systematic, documented, periodic and objective evaluation of the performance of the organisation, management system and processes designed to protect the environment with the aim of:*

- *facilitating management control of practices which may have impact on the environment*;
- *assessing compliance with company environmental policies.*

In order to participate in the scheme, a company must conduct the following continuous cycle of activities:

(1) Formulate and adopt a company environmental policy, taking into account:
   - compliance with legislation
   - reducing environmental impact to levels not exceeding those corresponding to the economically viable application of *best available technology*

(2) Conduct an environmental review of the site, covering:
   - energy management—fuel choice and savings
   - water management and savings
   - raw materials management—choice, transportation and savings
   - waste avoidance—reclaim, reuse, recycling, transportation and disposal; evaluation, control and reduction of noise
   - selection of new production processes and changes to existing processes
   - product planning (design, packaging, transportation, use and disposal)
   - environmental performance and practices of contractors, subcontractors and suppliers

- prevention and limitation of environmental accidents, together with contingency procedures
- staff information and training

(3) Construct an environmental statement for dissemination externally.
(4) Introduce an environmental programme for the site.
(5) Introduce an environmental management system for the site, aimed at achieving the commitments contained in the company environmental policy. This should include:
- environmental policy, objectives and programmes
- a definition of responsibilities and authorities of personnel
- training arrangements and needs identification
- procedures for documenting the activity
- environmental effects evaluation and registration. This should include registers of:
  - all legislative, regulatory and other policy requirements relating to the environment
  - controlled and uncontrolled emissions to atmosphere
  - controlled and uncontrolled discharges to water or sewers
  - solid and other wastes: particularly hazardous wastes
  - contamination of land
  - use of land, water, fuels and energy, and other natural resources
  - discharges of thermal energy, noise, odour, dust, vibration and visual impact
  - effects on ecosystems
- monitoring and control systems

(6) Carry out periodic *environmental audits*
(7) Set objectives for continuous *improvements*
(8) Go to Step 1—i.e. revise policy and continue to maintain the cycle.

The whole procedure must be validated by an independent external accredited environmental verifier. If successful, the company will be included in the eco-audit register of companies and will be entitled to use an eco-management and audit logo in relation to all its participating sites.

3.2 ENVIRONMENTAL MANAGEMENT SYSTEMS: BS 7750

The British Standards Institution (BSI) has recently produced a *Specification for Environmental Management Systems—BS 7750:1992* (2) (Figure 3.2), similar to BS 5750:1987 ((3) which deals with quality systems), with which it is analogous, and to which it has many links. BS 7750 also links to the EC Eco-Audit Regulation (1). The standard is applicable to any organisation which wishes to assure itself of

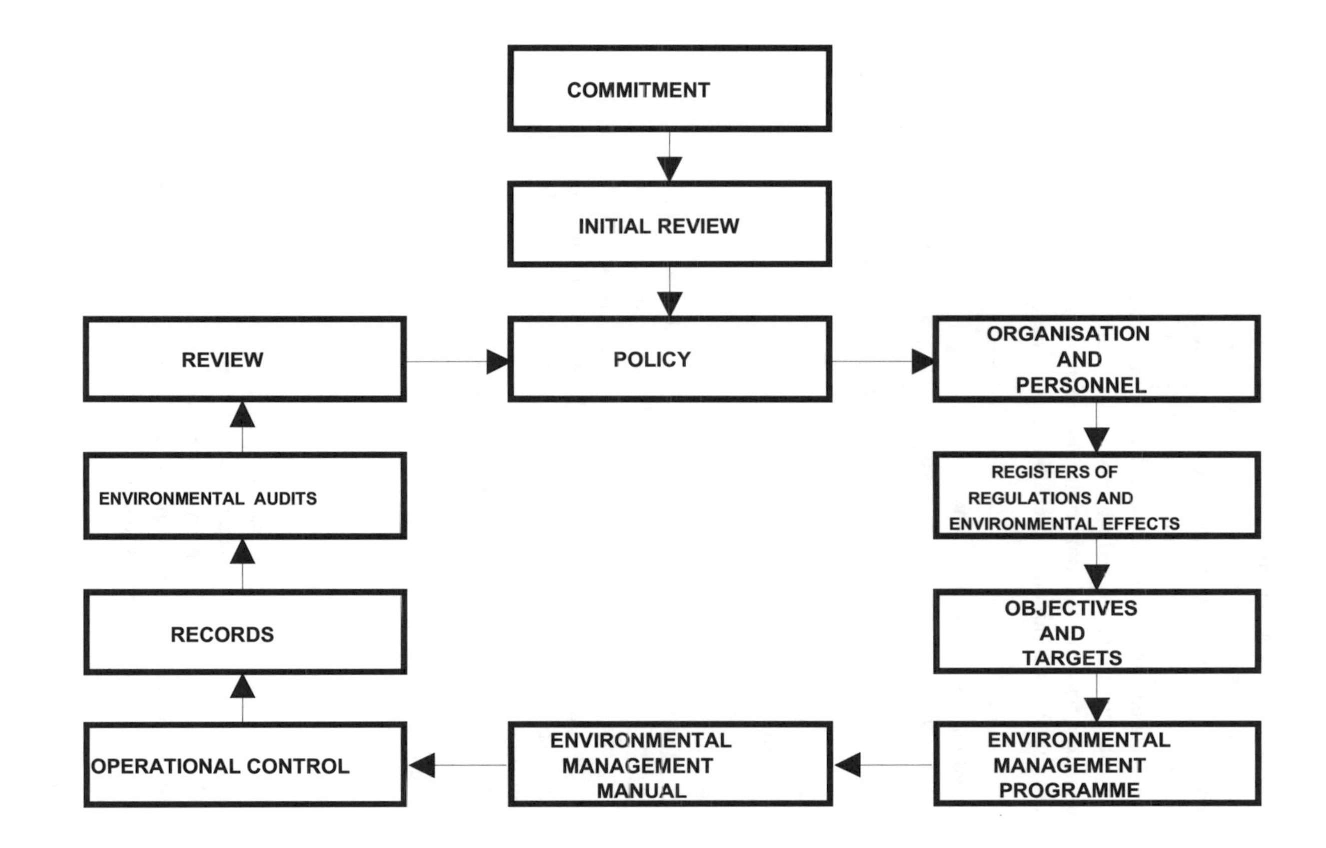

Figure 3.2 BS 7750

compliance with a stated environmental policy and demonstrate such compliance to others.

BS 7750 contains a specification for an environmental management system for ensuring and demonstrating compliance with stated environmental management policies and objectives. It is designed to enable an organisation to establish an effective cyclic environmental management system, as follows.

After a statement of commitment has been produced and an initial review of company objectives, e.g. to reduce waste, save materials, fuel and energy, to reduce or eliminate environmental pollution, and of past and current environmental performance has been conducted, the following 12 stages are ongoing and circulatory:

(1) Set *environmental policy.*
(2) *Educate* management and personnel.
(3) *Maintain records* of all legislative, regulatory and other requirements pertaining to environmental considerations. These may include planning conditions, discharge consents, process authorisations and improvement notices.
(4) Evaluate *environmental effects* considering:
    - emissions to the atmosphere
    - discharges to water
    - solid and other wastes
    - contamination of land
    - use of land, water, fuels and energy and other natural resources
    - noise, odour, dust, vibration and visual impact
    - effects on specific parts of the environment and ecosystems

    This should include effects arising from normal and abnormal conditions, incidents, accidents and emergencies, and should include past, current and planned activities.
(5) Set *environmental objectives and targets.*
(6) Establish and maintain an *environmental management programme* for achieving these objectives.
(7) *Maintain a manual*, recording policy, objectives and management programme.
(8) Set up *monitoring and control systems* to ensure that procedures are effectively followed.
(9) *Maintain records* to demonstrate compliance with the requirements and goals of the environmental management system.
(10) Carry out *audits* to determine whether environmental activities conform to the set management programme, and whether this programme is effective.
(11) *Review* the management system, results and policy.
(12) Go to Step 1.

Annexes A and B in BS 7750 suggest that the environmental probity of suppliers (of products and services) should also be considered with regard to their activities, products, packaging and delivery (cf. BS 5750).

3.3 ECOLABELLING

The process of establishing a framework for an award scheme of ecolabels for products was initiated by the European Commission by informing the European Parliament of the intention in 1987. In 1991, the EC commissioned pilot studies for the environmental labelling (or ecolabelling) of items to help prevent manufacturers from making ambiguous statements as to the 'environmentally friendly' nature of their products. The Ministers of the EC Environment Council adopted an **Ecolabelling Regulation 1992** (4), intended to enforce the introduction of such a scheme throughout the EU. The objectives are to:

- promote the design, production, marketing and use of products which have a reduced environmental impact during their entire life cycle
- provide consumers with better information on the environmental impact of products

At present, participation in the scheme and application for a label is voluntary.

The basis of the product ecolabel is the need to construct a *product environmental audit*. The system adopted should take into account the *full cradle-to-grave environmental impacts* of the product.

Life cycle analysis has been defined by the European Committee for Standardisation (reported in (5)) *as a method used to quantify environmental burdens based on an inventory of environmental factors for a product, process or activity from the abstraction of raw materials to their final disposal.*

A life cycle analysis involves examination of the inputs needed and outputs created, and must include in an inventory details of the energy, materials and water consumed, and the pollution of air, water and land produced, in the production, distribution (including packaging), utilisation and ultimate disposal of the product. Table 3.1 provides a matrix of operations and environmental effects which may be of use in constructing life cycle analyses of products. The matrix should be applied to both the product and its packaging, as well as imported energy, water and materials, and should include support services, maintenance requirements and their environmental impacts.

A *life cycle assessment* goes further than this by evaluating the effects of the life cycle analysis inventory on the environment as a whole, e.g. its contribution to global warming, ozone layer depletion, acid rain formation, the despoliation of water and land, as well as the knock-on effects of these global environmental problems. As a result, current attempts at life cycle assessment of products tend to be highly subjective and susceptible to manipulation depending upon the perceptions and attitudes of those conducting them.

BS 7750 states in Annex A that a company may wish to include in its environmentally related objectives to design products in such a way as to minimise their environmental effects in production, use and disposal. The evaluation of these

Table 3.1 Matrix of operations and environmental effects

Environmental Impact	Manufacture	Distribution	Utilisation	Disposal
Inputs				
Energy				
Water				
Materials				
Outputs				
Emissions to air				
Emissions to water				
Waste to land				
Noise				

effects should address all phases, through research and development, design, marketing, raw material sourcing, purchasing, production, waste management, packaging, storage, distribution, sales and use, to ultimate disposal.

Both direct effects (e.g. disposal/release of solid, liquid and gaseous wastes, energy and materials consumption during manufacture, the effects of transport, land management) and indirect effects (e.g. the extraction of fuels and other raw materials supplied by another organisation, and the ultimate disposal of the products from the system) should be included.

Clearly, the accurate quantification of such a label is difficult and, as such, currently remains to be performed voluntarily at the discretion of manufacturers to aid marketing. The possession of an approved ecolabel is, however, a positive statement regarding the environmental performance of a product relative to its competitors. This competitive advantage may be used to increase the price of the product or its market share.

REFERENCES

(1) EC Council Regulation allowing voluntary participation by companies in the industrial sector in a Community eco-management and audit scheme, Council Regulation (EEC) No. 1836/93, *Official Journal of the European Communities*, No. L 168/1, 10 July 93.

(2) *BS 7750, Specification for Environmental Management Systems*, British Standards Institution, 1992.
(3) *BS 5750, Specification for design, development, production, installation and servicing Quality Systems*, British Standards Institution, 1987.
(4) EC Regulation on Eco-labelling, No.880/92, 23 March 1992, *Official Journal of the European Communities*, L99, Vol. 35, 11 April 1992.
(5) N. Kirkpatrick, *Life Cycle Analysis and Eco-labelling*, PIRA International, Randalls Road, Leatherhead, Surrey, January 1992.

4

The integrated approach to environmental management: 12 steps to heaven

This chapter shows how integrated environmental management integrates the requirements of environmental legislation, the EC Eco-audit Regulation, BS 7750, the need to conserve energy, water and materials, the need to reduce polluting emissions to air, effluents to water and solid wastes to land, and the need to save money in the process, via reduced resource and utility costs, reduced effluent and waste disposal charges, reduced pollution taxes, the avoidance of fines and legal penalties and the avoidance of clean-up costs. The systematic application of integrated environmental management allows all necessary retrofits and modifications to technical activities to be financially resourced from within the environmental management project itself. The cyclic chain of activities to be conducted involves: education, commitment, an initial review, policy, organisation, compliance, a scoping study, management, documentation, a full review, the identification of options leading to recommendations, audits and reportage, concerning energy and emissions to air, water and effluents, waste on land, materials, products and packaging, transport and noise. This cycle is developed and fleshed out here.

4.1 INTEGRATED ENVIRONMENTAL MANAGEMENT

The integrated approach to environmental management (see Table 4.1 and Figure 4.1) integrates the requirements of:

- environmental legislation
- the EC Eco-audit Regulation
- BS 7750
- the need to conserve energy, water and materials

Table 4.1 Integrated environmental management

Environmental management	Environmental engineering
Problems	Energy
Legislation	Water
Standards	Materials
Costs and Taxes	Waste
Policy	Noise
Impact	Products

- the need to reduce polluting emissions to air, effluents to water and solid wastes to land
- the need to save money in the process, via:
  - reduced resource and utility costs
  - reduced effluent and waste disposal charges
  - reduced pollution taxes
  - the avoidance of fines and legal penalties
  - the avoidance of clean-up costs

For the majority of companies, the systematic application of integrated environmental management allows all necessary retrofits and modifications to technical activities to be resourced financially from within the environmental management project itself.

The systematic management of energy and materials resources accomplishes:

Wants

- pure air
- pure water
- pure land
- pure food
- pure animals
- pure fish
- conservation of resources

Don't wants

- pollution
- poison
- unnatural products
- waste

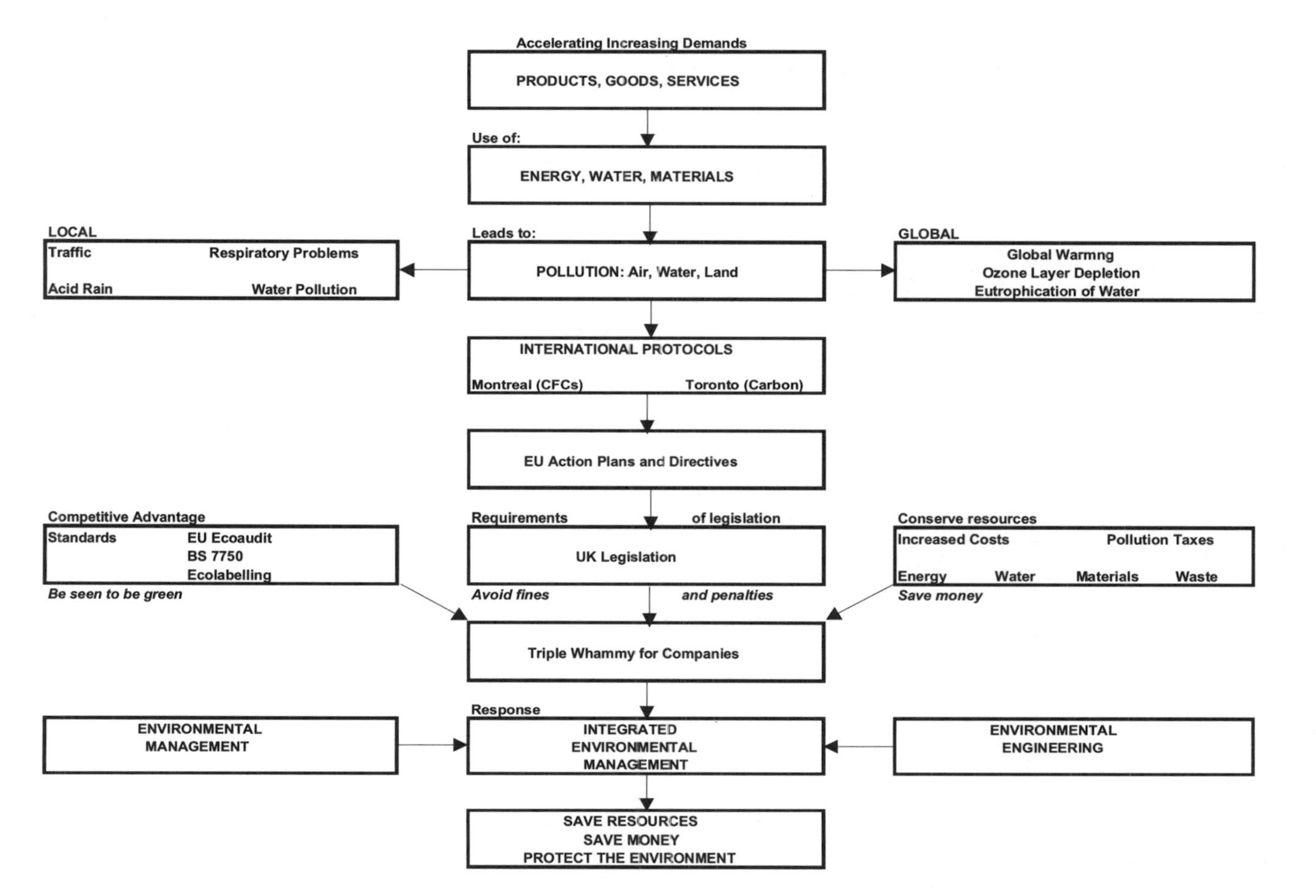

Figure 4.1 The birth of integrated environmental management

Table 4.2 Integrated environmental management disciplines

Energy	Water	Materials
Flue gases	Effluents	Waste
Air pollution	Water pollution	Land pollution
Flue gas clean-up	Water clean-up	Waste handling
Energy management	Water management	Materials management

Environmental management (2) comprises those *aspects of the overall management function (including planning) that determine and implement the company's environmental policy*. It considers local and global environmental problems, relevant legislation, standards, costs and taxes and overall environmental impact. It concerns the formulation and control of environmental policies and management systems with respect to energy, water, materials and waste.

Environmental engineering *is a technical function which implements the environmental policies*. Its remit is to monitor, record, analyse, critically examine, alter and control energy and materials flows through systems so that resources are utilised with maximum efficiency. It involves *environmental accounting*, which provides the numerical information needed to quantify the system's degree of imbalance with the environment, aiding management decision making and so leading to recommendations for technical action.

It embraces the disciplines of engineering, science, mathematics, economics, accountancy, design and operational research, computation and information technology. The environmental engineer must also be responsible for the day-to-day management of materials and water, fuels, furnaces, boiler houses, distribution systems, building services, plant, process equipment, polluting exhausts, effluents and waste.

Integrated environmental management thus includes **energy management, water management, materials management** and **waste management** of sites, buildings, services, processes and products (see Table 4.2).

4.2 ENVIRONMENTAL ISSUES

Environmental issues of concern come under the following general headings, which are expanded on below.

- General issues
- Processes
- Substances
- Services
- Transportation

- Energy
- Emissions to air
- Water
- Discharges to water
- Materials
- Waste
- Other nuisances
- Nature and ecology
- Accidents, health and safety
- Products
- Packaging
- Information

General issues

- Strengths of the organisation
- Weaknesses of the organisation
- Environmental, health, safety, legal and financial risks engendered
- Areas for concern
- Legal considerations
- Licences needed
- Adherence to standards

Site description

- Land system boundaries
- Inventory of land involved
- History
- Historical review of the site
- Historical uses for land
- Environmental hazards from past uses
- Contaminated ground or groundwater
- Environmental impacts
- Current uses of land
- Present operations
- Maintenance programmes

Processes

- Selection of new processes and alterations to existing processes
- Assessing impacts of new activities in advance

- Minimising the environmental effects of new activities by advanced planning
- Descriptions of processes
- Functions of processes
- Compliance with legislation and regulations
- Scheduled processes which are covered by HMIP
- Scheduled processes which are covered by Local Authority controls
- Scheduled processes covered in Part A of the IPC regulations of the Environmental Protection Act
- Scheduled processes covered in Part B of the IPC regulations of the Environmental Protection Act
- Impacts on the local environment
- Impacts on the global environment
- Minimising waste of energy, water and materials by process improvement
- Measures necessary to minimise, prevent or eliminate pollution
- Monitoring
- Measurements
- Testing
- Controls

Substances

- Types
- Inventory
- Sources
- Quantities
- Qualities
- Uses
- Frequencies of use
- Costs
- Stocks
- Hazardous substances and environmental impact
- Monitoring
- Controls
- Transport

Services

- Buildings inventory
- Facilities inventory
- Current usage
- Past usage

- Current environmental effects of past usage
- Present environmental hazards (e.g. asbestos insulation)
- Building services:
  - heating
  - ventilating
  - lighting
  - refrigerating
  - air conditioning
  - hot water services
  - compressed air services
  - waste disposal systems: to air, to water, to land

Transportation

- Traffic pollution produced in transportation to, within and from the site
- Energy consumed in transportation to, within and from the site
- Noise produced in transportation to, within and from the site
- Air pollution produced in transportation to, within and from the site
- Formulation of an environmentally sound transport policy: vehicles, fuels, routes and time schedules

Energy: inventory and energy accounts

Imports

- Energy choice and types
- Sources
- Quantities
- Qualities
- Consumption figures
- Fuels and electricity costs
- Stocks
- Fuels and electricity delivered:
  - coal
  - oil
  - gas
  - electricity
  - other
- Uses of energy (fuels and electricity):
  - heating
  - ventilating

  - lighting
  - refrigerating
  - air conditioning
  - hot water
  - compressed air
  - other
- Energy flows
- Energy management
- Energy rejection to the environment
- Conservation and energy savings:
  - use of renewable energy
  - use of clean technologies (e.g. combined heat and power systems)
  - waste avoidance
  - reduction of energy use
  - energy cascading
  - thermal insulation
  - heat recovery
  - waste-to-energy options
  - monitoring
  - controls

Exports

Emissions to air

- Complaints received
- Controlled and uncontrolled emissions to air
- Energy-related emissions:
  - types
  - sources
  - quantities
  - qualities
  - frequencies of emissions
  - dispersion
  - monitoring
  - environmental impact
  - complaints received
  - need for flue gas clean-up
  - reduction of emissions
  - discharge control
  - rendering harmless unavoidable emissions
  - costs

- Other emissions to air:
  - types
  - substances
  - sources
  - quantities
  - qualities
  - frequencies of emissions
  - dispersion
  - monitoring
  - environmental impact
  - complaints received
  - needs for clean-up
  - reduction of emissions
  - discharge control
  - rendering harmless unavoidable emissions
  - costs

Water: inventory and water accounts

Imports

- Water delivered:
  - supplies
  - sources
  - quantities
  - qualities
- Water consumption
- Costs for water
- Uses for water
- Water flows
- Water wastage
- Water management:
  - reduction of water use
  - waste water avoidance
  - water conservation and savings
  - use of clean technologies
  - use of alternative water sources
  - rainwater harvesting
  - stormwater
  - groundwater
  - reuse of water
  - water cascading

- reclamation of water
- water recycling
- monitoring
- controls

Exports

Discharges to water

- Complaints received
- Inventory of discharges to water
- Inventory of effluents
- Energy-related discharges to water
- Other controlled and uncontrolled discharges of wastes to water
- Controlled and uncontrolled effluents
- Sinks for wastes to water and effluents
- Liquid wastes (hazardous or other) and environmental impact
- discharge monitoring
- Water sampling:
  - quantities
  - qualities
  - frequencies of discharges
- Discharge control
- Reduction of effluents
- Effluent clean-up
- Needs for water clean-up
- Costs

Materials

Imports

- Raw materials choice
- Stocks
- Storage arrangements
- Environmental impact of raw material sourcing
- Controlling the environmental effects of raw material sourcing
- Inventory and materials accounts
- Materials delivered:
  - grades
  - sources

  - quantities
  - qualities
  - consumptions
  - costs
- Uses of materials
- Materials flows
- Materials management:
  - conservation and savings
  - waste avoidance
  - reduction of raw materials uses
  - use of renewable materials
  - use of clean technologies
  - materials cascading
  - reclaim
  - reuse
  - recycle
  - monitoring
  - controls
- Transportation of materials
- Toxic materials
- Radioactive materials
- Environmental impact

Exports

Waste

- Inventory of wastes
- Sources of waste
- Analyses of waste:
  - types
  - energy-related solid wastes
  - other controlled and uncontrolled solid waste discharges to land: quantities, qualities, frequencies of discharges
- Sinks for waste
- On-site waste disposal methods:
  - disposal to land: degradation
  - disposal to water: dilution
  - disposal to air: incineration
- Contamination of land
- Environmental impacts
- Complaints received

- On-site waste management:
  - reduction of waste quantities
  - minimisation of waste quantities
  - elimination of waste
  - reclaim
  - reuse
  - recycle
  - waste-to-energy options
  - monitoring
  - discharge control
  - rendering harmless unavoidable wastes
- Waste disposal:
  - costs
  - precautions for the storage, handling and transport of dangerous waste
  - waste transportation
  - ultimate waste disposal
  - environmental impacts

Other nuisances

- Thermal energy
- Noise:
  - sources of noise
  - acoustical protection requirements
  - measurements, evaluation, control and reduction of noise pollution
- Vibration:
  - complaints received
  - sources of vibration
  - evaluation, control and reduction of vibration
- Dust:
  - sources of dust
  - evaluation, control and reduction of dust production and release
- Odours:
  - sources of odours
  - evaluation, control and reduction of odour production and release
- Monitoring
- Controls

Nature and ecology

- Effects on ecosystems
- Nature conservation

- Litter
- Visual impact
- Landscaping
- Trees
- Plants
- Wildlife
- Protecting local ecology
- Protecting global ecology

Accidents, health and safety

- Prevention and mitigation of accidents
- Measures necessary to prevent accidental emissions of energy or materials
- Health and safety considerations and arrangements on site
- Control of special substances hazardous to health (COSHH register) and prescribed substances
- Procedures for handling special and prescribed substances
- Spill control measures
- Fire control measures
- Systems and procedures for dealing with accidents and emergencies:
  - internal
  - external
- Monitoring
- Controls
- Systems for dealing with the local community, press and media

Products

Imports

- Inventory of imported products
- Uses of imported products
- Reduction, minimisation or elimination of imported products

Exports

- Inventory of exported products
- Product planning
- Product life cycles: cradle-to-grave environmental impact assessments
- Designing products so as to minimise their overall cradle-to-grave environmental impact

- Waste avoidance in product manufacture
- Product transportation
- Use of exported products
- Energy consumption of exported products
- Water consumption of exported products
- Materials consumption of exported products
- Air pollution by exported products
- Water pollution by exported products
- Waste production by exported products
- Reclamation of products
- Reuse of products
- Recycling of products
- Ultimate disposal of products

Packaging

Imports

- Paper and packaging policies
- Packaging requirements: materials, quantities and qualities used
- Packaging of imports
- Inventory of imported packaging
- Use of imported packaging
- Need for imported packaging
- Disposal of imported packaging
- Recyclability of imported packaging
- Recycling of imported packaging

Exports

- Paper and packaging policies
- Exported packaging of products
- Packaging requirements: materials, quantities and qualities used
- Transportation of exported packaging
- Reclamation of exported packaging
- Reuse of exported packaging
- Recyclability of packaging
- Recycling of exported packaging
- Ultimate disposal of product packaging

Information

- Education
- Staff training
- Skills
- Records
- Registers
- Internal information
- Staff participation
- External information
- Environmental performance and practices of contractors, subcontractors and suppliers

4.3 DEFINITIONS

Auditor (1). *An individual or a team, belonging to a company personnel or external to the company, acting on behalf of company top management*, possessing, individually or collectively, competencies as follows:

- Knowledge of the:
  — sectors audited
  — fields audited
- Knowledge and experience of the relevant:
  — environmental management issues
  — technical issues
  — environmental issues
  — regulatory issues
- Sufficient training and proficiency in the specific skills of auditing to achieve the stated objectives.
- Sufficient independence from the activities audited to be able to make objective and impartial judgements.

Audit cycle (1). *The period of time in which all the activities in a given site are audited.*

Environment (2). *The surroundings and conditions in which an organisation operates, including living systems (human and other) therein.* As the environmental effects of the organisation may reach all parts of the world, the environment in this context extends from within the workplace to the global system.
Environment (3). *All of the surroundings of an organism, including other living things, climate and soil, etc.* In other words, the conditions for development or growth.

Environmental account. *A balance sheet of energy and materials INPUTS, OUTPUTS and THROUGHPUTS concerning the system under scrutiny.* It seeks to quantify all inputs, throughputs and outputs to, within and from a system boundary in terms of past, present and future environmental impacts.

Environmental audit (1). *A management tool comprising a systematic, documented, periodic and objective evaluation of the performance of the organisation, management system and processes designed to protect the environment with the aim of:*

- *facilitating management control of practices which may have impact on the environment*
- *assessing compliance with company environmental policies*

(See also environmental management audit.)

Environmental audit (3). *An accounting by manufacturers of the products produced and their effects on the environment*—energy use policies, materials use policies, waste outputs and their effects on the environment. Purchasing procedure monitoring to obtain environmentally sound goods and services, etc. may also be adopted as part of an environmental audit.

Environmental audit (4). *A systematic and comprehensive analysis of the company's impact on the environment.*

Environmental auditing (5). *A series of activities, initiated by management, to evaluate environmental performance, to check compliance with environmental legislation and to assess whether the systems in place to manage environmental improvements are effective.* Audits are done at regular intervals to assess the environmental performance of the company in relation to the company's stated objectives and environmental policy.

Environmental auditing (1). The process of monitoring and targeting against the baseline to assess adherence to:

- policy
- objectives
- targets
- plans
- strategies
- standards set by legislation and regulations
- best practice elsewhere

Environmental auditing (British Association of Environmental Consultancies). *A full environmental audit comprises a complete examination of all aspects of a company's interactions with the environment.* It is designed to examine performance in terms of:

- management
- compliance with policies and regulations
- information to be disclosed to the authorities or to the public

The full environmental auditing procedure, as proposed by the European Commission, is outlined below.

The audit should include at least the following steps, which should be documented:

(1) Planning of the audit activities, including a definition of responsibilities.
(2) Review of the environmental protection policies of the company.
(3) Assessment of the organisation, its management, and its equipment.
(4) Gathering of data and all relevant information.
(5) Evaluation of the overall performance.
(6) Identification of areas for improvement.
(7) Internal reporting to top management.

A follow-up policy statement by top management should conclude the auditing procedure. At least the following aspects should be considered by the audit:

- Relevant information and data on material and energy exchanges with the environment due to the activity being audited.
- Compliance with community, national and local regulations and standards applicable to the establishments concerned.
- Significant incidents or operation disturbances which have occurred, or complaints received.
- A review of the available policy, equipment, organisation and management systems related to the following aspects:
  - pollution discharge control, monitoring and reduction
  - other nuisances
  - energy choice and reduction of energy use
  - raw materials choice and transportation: reductions of water and raw materials uses
  - waste transportation, elimination, recycling and reuse
  - product planning (design, packaging, transportation, use and elimination)
  - prevention and mitigation of accidents
  - staff training and participation
  - external information

Environmental compliance audit (5). This checks the extent to which an organisation is complying with existing environmental laws and company policies.

Environmental effect (2). *Any direct or indirect impingement of the activities, products and services of the organisation upon the environment, whether adverse or beneficial.*

Environmental effects evaluation (2). *A documented evaluation of the environmental significance of the effects of the organisation's activities, products and services (existing and planned) upon the environment.*

Environmental effects register (2). *A list of the significant environmental effects,* *known or suspected, of the activities, products and services of the organisation upon the environment.*

Environmental engineering is *a technical function, the remit of which is to monitor, record, analyse, critically examine, alter and control energy and materials flows through systems so that resources are utilised with maximum efficiency.*

Environmental impact assessment (EIA) (3). This is necessary to ensure that new developments take into account environmental considerations at the planning stage. The results of the EIA are embodied in the **environmental impact statement** (EIS) (3), which is now often required by law, and which should contain the following information:

- a description of the development proposed, the site and the design and size or scale of the development
- the data necessary to identify and assess the main effects that the development is likely to have on the environment
- a description of the likely significant direct and indirect effects of the development by reference to its possible impacts on human beings, flora, fauna, soil, water, air, climate, landscape, cultural heritage, employment, transport, education resources, housing, etc.
- where significant adverse effects are identified, a description of the measures envisaged in order to avoid, reduce or remedy these
- a summary in non-technical language for public consumption

Environmental management (2). *Those aspects of the overall management function (including planning) that determine and implement the environmental policy.*

Environmental management audit (2). *A systematic evaluation to determine whether or not the environmental management system and environmental performance comply with planned arrangements, and whether or not the system is implemented effectively, and is suitable to fulfil the organisation's environmental policy.* (See also environmental audit.) Environmental management audits usually review environmental management systems, focusing on the development of policy, appropriate procedures and systems, and product life cycle (cradle-to-grave) analytical techniques.

Environmental management manual (2). *The documentation describing the procedures for implementing the organisation's environmental programme.*

Environmental management programme (2). *A description of the means of achieving environmental objectives and targets.* (See also environmental programme.)

Environmental management review (2). *The formal evaluation by management of the status and adequacy of systems and procedures in relation to environmental issues, policy and regulations as well as new objectives resulting from changing circumstances.* (See also environmental review.)

Environmental management system (1). *That part of the overall management system which includes the organisational structure, responsibilities, practices, procedures, processes and resources for determining and implementing the environmental policy.*
Environmental management system (2). *The organisational structure, responsibilities, practices, procedures, processes and resources for implementing environmental management.*

Environmental objectives (1). *The detailed goals, in terms of environmental performance, which a company sets itself.*
Environmental objectives (2). *The goals, in terms of environmental performance, which an organisation sets itself to achieve and which should be quantified wherever practicable.*

Environmental policy (1). *A company's overall aims and principles of action with respect to the environment including compliance with all relevant regulatory requirements regarding the environment.*
Environmental policy (2). A public statement of the intentions and principles of action of the organisation regarding its environmental effects, giving rise to its objectives and targets.

Environmental product audit (5). *The analysis of a particular product line examining most aspects of sourcing production, packaging and waste disposal.* This is in effect a life cycle assessment of the product.

Environmental programme (1). *A description of a company's specific objectives and activities to ensure greater protection of the environment at a given site, including a description of the measures taken or envisaged to achieve such objectives and where appropriate the deadlines set for implementation of such measures.* (See also environmental management programme.)

Environmental review (1). *A comprehensive analysis of the environmental issues, impact and performance related to activities at a site.* (See also environmental management review.)

Environmental targets (2). Detailed performance requirements, quantified wherever practicable, applicable to the organisation or parts thereof, which arise from the environmental objectives and which need to be met in order to achieve these objectives.

Environmental statement (1). *A statement prepared by the company in line with the requirement of the EC Eco-audit regulation*, which should include

- a description of the company's activities at the site considered
- an assessment of all the significant environmental issues relevant to these activities

- a summary of the figures on energy consumption, water consumption, raw materials consumption, pollutant emissions, waste generation, and other significant environmental aspects
- the company's environmental policy, objectives and targets
- details of the programme to be followed and the environmental management system

Integrated environmental management. *This integrates environmental management with environmental engineering*, and so covers environmental problems, legislation, standards, economics, policy and impact with respect to energy and air pollution, water and effluents, the use of materials, the production of waste, the production of land pollution, noise abatement, products and ecolabelling.

Liability audits *are most often associated with mergers and acquisitions and are designed to assess potential liabilities in relation to clean up costs, taking account of current and future legislation.*

Register of regulations (2). *A register of legislative, regulatory, codes of practice and other policy requirements.*

4.4 ESSENTIAL REFERENCE MATERIAL

British Standards Institution, *Specification for Environmental Management Systems, BS 7750*, 1992.

British Standards Institution, *Quality Systems, BS 5750*, 1987.

Department of the Environment and the Welsh Office, *Integrated Pollution Control: A Practical Guide*, HMSO, London, 1993.

EC Council Regulation allowing voluntary participation by companies in the industrial sector in a Community eco-management and audit scheme, Council Regulation (EEC) No. 1836/93, *Official Journal of the European Communities*, No. L 168/1, 10.7.93.

EC Eco-Audit Regulation (Draft Version 3, Dec. 1991).

Environmental Protection Act, HMSO, 1990.

Environmental Protection (Prescribed Processes and Substances) Regulations, HMSO, 1991.

Environmental Protection (Applications, Appeals and Registers) Regulations, HMSO, 1991.

Environmental Protection (Authorisation of Processes) (Determination Periods) Order, HMSO, 1991.

Her Majesty's Inspectorate of Pollution, *Chief Inspector's Guidance to Inspectors, Environmental Protection Act 1990, Process Guidance Note IPR: Fuel and Power*, HMSO, London, 1991.

Her Majesty's Inspectorate of Pollution, *Chief Inspector's Guidance to Inspectors, Environmental Protection Act 1990, Process Guidance Note IPR 2: Metal*, HMSO, London, 1991.

Her Majesty's Inspectorate of Pollution, *Chief Inspector's Guidance to Inspectors, Environmental Protection Act 1990, Process Guidance Note IPR 3: Mineral*, HMSO, London, 1991.

Her Majesty's Inspectorate of Pollution, *Chief Inspector's Guidance to Inspectors, Environmental Protection Act 1990, Process Guidance Note IPR 4: Chemical*, HMSO, London, 1991.

Her Majesty's Inspectorate of Pollution, *Chief Inspector's Guidance to Inspectors, Environmental Protection Act 1990, Process Guidance Note IPR 5: Waste Disposal*, HMSO, London, 1991.
Her Majesty's Inspectorate of Pollution, *Chief Inspector's Guidance to Inspectors, Environmental Protection Act 1990, Process Guidance Note IPR 6: Other Industries*, HMSO, London, 1991.
Water Act, HMSO, 1989.
Water Industry Act, HMSO, 1991.
Water Resources Act, HMSO, 1991.

Reference material should also include documentation on suppliers, and an inventory of all incoming supplies, raw materials and products:

- nature and composition
- conditions
- quantities
- qualities
- environmental impact
- health effects

4.5 THE CHAIN: AN APPROACH TO INTEGRATED ENVIRONMENTAL MANAGEMENT

The EC Eco-audit Regulation states that industry has its own responsibility to manage the environmental impact of its activities and should therefore adopt a proactive approach in this field.

When a company wishes to consider environmentally related matters as part of its overall remit, in order to introduce the necessary procedures economically and with utmost efficiency, it should attempt to satisfy simultaneously the requirements of environmental legislation, the EC Eco-audit Regulation, BS 7750, the need to conserve energy, water and materials, the need to reduce polluting emissions to air, effluents to water and solid wastes to land, and the need to save money in the process, via reduced resource and utility costs, reduced effluent and waste disposal charges, reduced pollution taxes, and the avoidance of fines and legal penalties and clean-up costs.

If these requirements are addressed simultaneously, all necessary retrofits and modifications to technical activities, dictated by the legislation and the standards, may be financially resourced from savings achieved within the environmental management project itself, via reduced energy, water and materials costs, reduced charges for effluent and solid wastes disposal, and the avoidance or minimisation of pollution taxes, fines and legal penalties and clean-up costs.

In order to accomplish this, the company must institute and follow a chain of actions as listed in Table 4.3 and amplified in Table 4.4.

Table 4.3 The chain: 12 steps to heaven

1	Education	Understand global, national and local environmental *problems and issues*
2	Commitment	Make a *company commitment* to environmental responsibility
3	Initial review	Conduct an initial environmental review
4	Policy	Set/revise company *environmental policy* and programmes Make *environmental statement*
5	Organisation	Structure/restructure the *environmental organisation*
6	Compliance	Assess and ensure *compliance with legislation and regulations* in both environmental and health and safety areas
7	Scoping study	Perform an *environmental scoping study* Set *objectives and targets*
8	Management	Set up/revise *environmental management system*
9	Documentation	Establish/revise *documentation systems*
10	Review	Conduct *environmental review*
11	Options/ recommendations	Identify and evaluate *options for improvements*
12	Audit/report	Conduct *environmental audit* and report

Go to Step 4 and keep pedalling.

Table 4.4 The matrix

No.	**Step**	**Energy and emissions to air**	**Water and effluents**	**Waste on land**	**Materials, products and packaging**	**Transport**	**Noise health and safety**
1	Education	×	×	×	×	×	×
2	Commitment	×	×	×	×	×	×
3	Initial review	×	×	×	×	×	×
4	Policy	×	×	×	×	×	×
5	Organisation	×	×	×	×	×	×
6	Compliance	×	×	×	×	×	×
7	Scoping study	×	×	×	×	×	×
8	Management	×	×	×	×	×	×
9	Documentation	×	×	×	×	×	×
10	Review	×	×	×	×	×	×
11	Options/ recommendations	×	×	×	×	×	×
12	Audit/report	×	×	×	×	×	×

This chain of activities is cyclic and continuous. The initial progression around the chain will, of necessity, be fast and loose, so that immediate problems, regarding in particular legislative and health and safely aspects of the organisation, may be identified. As the cycle continues, the studies will be conducted less rapidly and in more depth, leading to greater efficiency as the process continues.

Step 1: Education—understand global, national and local environmental problems and issues

The EC Eco-audit Regulation (1) states that the application of environmental management systems by companies should take account of the need to ensure awareness and training of workers in the establishment, especially those personnel whose work may have a significant effect on the environment.

BS 7750 requires that the organisation should ensure that its members are aware of the importance and benefits of compliance with its environmental policy and objectives, as well as individual roles in achieving this compliance.

Needs must be identified and education and training programmes set up for management and workers. Subject areas to be covered should include:

- global environmental problems: the greenhouse effect, global warming, climate change, ozone layer depletion, water pollution and its effects, resource depletion and conservation
- local environmental problems: traffic pollution from road vehicles, roads and transportation infrastructure, air pollution from vehicles and its effects, air pollution from power generation and factories, water pollution from people, industry and agriculture, land pollution from increasing quantities of domestic and industrial waste, and the disposal of wastes
- the Toronto and Montreal Protocols and international and national reactions
- consumerism and conservationism
- Agenda 21 and sustainable development
- EC Environmental Action Programmes
- EC environmental Directives
- UK environmental policy
- UK environmental legislation and regulations
- prescribed processes, substances and BATNEEC
- EC and UK environmental standards and codes of practice
- principles and procedures of integrated environmental management

Information concerning the company's environmental activities must also be supplied to regulatory agencies, employees, customers, suppliers and trading partners, investors, shareholders and other stakeholders, insurers, banks and other lenders, environmental groups and interested parties, the media, the local community and the public at large.

This information should include:

- a statement of **company commitment** to environmental responsibility
- the **company environmental policy statement**
- a **statement of environmental objectives and targets**
- the **company environmental programme**
- the **company general environmental statement**
- the structure and responsibilities of the **environmental management team**
- the **register of regulations**
- a **statement of compliance with legislation and regulations** in both environmental and health and safety areas
- details of the **environmental management system**
- the report of the **environmental review**
- the **environmental effects register**
- the **inventory of local and global environmental effects**
- a list of the recommended **options for improvements**
- the **environmental audit report**
- **records of training**

Most of this information will be contained in the **environmental management manual**.

Documentation

Records of training

Step 2: Commitment—make a company commitment to environmental responsibility

The EC eco-audit Regulation (1) suggests that a company's environmental policy should include a statement of commitment aimed at the reasonable, continuous improvement of environmental performance, with a view to reducing environmental impacts to levels not exceeding those corresponding to *the economically viable application of best available technology*.

The company's environmental programme must form an intrinsic, high priority part of all management activities and cannot be conducted in a piecemeal or *ad hoc* manner. The entire organisation from the top management to the lowliest worker must share a **company commitment** to environmental management and improvement, which involves the whole organisation. Adequate resources and personnel should be assigned to the activity.

Thus the company must make a corporate commitment to environmental improvement and communicate this commitment to regulatory agencies, employees, customers, suppliers and trading partners, investors, shareholders and other stakeholders, insurers, banks and other lenders, environmental groups and interested parties, the media, the local community and the public at large.

Documentation

Statement of company commitment, for example:

> This company as a whole regards the safe-guarding of the environment as a high priority in its affairs and is dedicated to committing the necessary resources and personnel to achieve this aim, via complying with environmental legislation and accepted standards of performance, conserving energy, water and materials and reducing polluting emissions to air, effluents to water and solid wastes to land.

Step 3: Initial review—conduct an initial environmental review

BS 7750 requires that an organisation having no existing formal environmental management system should conduct an initial **review** to ascertain its current position with regard to the environment.

The aim of this *initial* review is to identify strengths, weaknesses, risks and opportunities as a basis for establishing the environmental management system. The nature and extent of problems and deficiencies should be identified, and priorities for action and resource needs formulated.

The following areas should be covered:

- environmental legislative and regulatory requirements
- health and safety regulations
- significant environmental effects
- procurement
- existing environmental management practices and procedures
- emergency procedures
- feedback mechanisms with regard to previous environmental incidences and non-compliance

This should be conducted via:

- questionnaires
- checklists
- interviews
- workshops
- inspection
- measurement

and should cover:

- processes and substances
- energy and emissions to air

- water and effluents
- waste on land
- waste disposal
- materials, especially hazardous materials
- products and packaging and their ultimate disposal
- recycling
- transport and noise
- nature conservation
- litter
- visual impact
- landscaping
- odours

as well as:

- services
- information
- records
- training
- health and safety matters

The list of **environmental issues** should be used as a checklist, but only rapid annual estimates of inputs, throughputs and outputs pertaining to energy and emissions to air, water and effluents, materials utilisation and the disposal of waste on land (masses, volumes, values) are necessary at this stage.

Documentation

The *initial* environmental review report.

Step 4: Policy—set/revise company environmental policy and programmes, make environmental statement

The EC Eco-audit Regulation (1) calls for companies to set and implement **environmental policies, objectives, programmes** and **environmental management systems** for improving environmental performance. Periodic **environmental statements** should be produced for dissemination to the public describing these. According to this regulation, *these statements should be validated by an external accredited environmental verifier.*

The policy should provide for compliance with all relevant regulatory requirements and include the statement of company commitment.

BS 7750 suggests that the organisation should define and document its environmental policy, which should:

- be relevant to its activities, products and services and their environmental effects
- be initiated, developed and actively supported by management at the highest level within the organisation
- be understood, implemented and maintained at all levels within the organisation
- commit the organisation to meet all relevant regulatory requirements
- seek to exceed these requirements
- include a commitment to continual improvement of environmental performance
- provide for the setting and publication of environmental objectives
- be consistent with health and safety policies
- be made publicly available

It should consider:

Principles

- Principles of sustainable development and ethical considerations
- Integrating environmental factors into all business decisions
- Requirements of suppliers
- Purchasing strategy
- Adopting an environmentally sound transport strategy
- Making a commitment to continuous improvement
- Long-term improvement goals
- Rate of improvement and how to measure it
- Impacts on nature and ecosystems
- Planting trees and vegetation

Information

- Fostering a sense of responsibility for the environment at all levels of employees
- Providing environmental education for company personnel
- The need for staff information
- Training needs and staff involvement
- Maintaining records
- Environmental information disclosure to suppliers, customers and the general public
- Advising customers on the relevant environmental aspects of the handling, use and disposal of its products

Regulations

Meeting and exceeding the requirements of legislation and accepting liabilities, regarding:

- emissions to air
- waste to water
- waste to land
- health and safety matters

conforming to standards and going beyond the minimum requirements. The highest available standards should be sought.

Commitments and expectations

- Making commitments to the local community and the public at large
- Assisting in local environmental initiatives
- Reviewing the practices of competitors
- External relations and giving wider support for environmental initiatives
- Expectation of high environmental standards from suppliers, contractors and vendors
- Involvement and integration of the supply chain

Environmental input/outputs

The components of environmental issues relating to:

- processes
- substances
- services
- transportation
- energy
- emissions to air
- water
- discharges to water
- materials
- waste
- other nuisances
- nature and ecology
- accidents, health and safety
- products
- packaging

should be considered when constructing the environmental policy.

Assessments and impacts

- Assessing and monitoring current impacts and aiming to lessen these
- The need for monitoring, measurements and testing

- Assessing impacts of new activities in advance
- Minimising the environmental effects of new activities by advance planning
- Impacts on the local environment
- Protecting local ecology
- Impacts on the global environment
- Protecting global ecology

Actions

- Future actions:
  - now
  - this year
  - in the future
- Implementing an environmental management system
- Constructing environmental reviews and audits
- Conducting annual reviews of policy, objectives and targets

Documentation

Policy statement on the environment and environmental statement.

Step 5: Organisation—structure/restructure the environmental organisation

BS 7750 (2) requires that the organisation should define and document the responsibility, authorities and interrelations of the key personnel who manage, perform and verify work affecting the environment, including those who need the organisational freedom and authority to:

- provide sufficient resources and personnel for implementation
- initiate necessary actions
- identify and record environmental problems
- recommend and initiate solutions
- verify implementation of the actions involved
- act in emergencies

The environmental management team must have sufficient independence from the activities audited to be able to make objective and impartial judgements.

Select a team leader and team who have technical training, expertise and knowledge of:

- the organisation
- the technical operations involved

- environmental issues
- environmental legislation and regulations
- environmental standards
- economics—costs and taxes
- auditing procedures and techniques
- accounting procedures
- environmental marketing

(See also scoping study and management.)

Documentation

Structure and responsibilities of the integrated environmental management team.

Step 6: Compliance—assess and ensure compliance with legislation and regulations in both environmental and health and safety areas

BS 7750 (2) requires that the organisation should establish a **register of regulations**: a register of legislative, regulatory and other policy requirements pertaining to the environmental aspects of its activities, products and services.

Check the requirements of LAAPC (Chapter 3).

Application for authorisation to emit under IPC regulations

The procedure is to:

- identify prescribed substances
- identify prescribed processes
- identify BATNEEC
- incorporate BATNEEC

The application must include:

- plant description
- an inventory of prescribed processes
- energy flows
- materials flows
- emissions to air—present and future plans
- emissions to water—present and future plans
- emissions to land—present and future plans
- an inventory of prescribed substances
- details of BATNEEC, BPEOs and monitoring facilities
- details of surroundings

(See also Chapter 3)

This information will be entered on a public register, together with the decision of HMIP.

Records must also be kept of incidents (e.g. accidents, emergencies, emissions above set standards, control device failures, leaks, spills, accidental discharges, customer complaints, public complaints, contractors' non-compliance with company environmental policy), failures to comply with policy and the remedial actions taken.

Documentation

Register of regulations.

Standards:

- EC Eco-Audit Scheme, BS 7750
- EC Action Programmes, Directives, UK Acts and Regulations:
  - — planning requirements
  - — health and safety (COSHH Register)
  - — use of energy, water and materials
  - — emissions, effluent discharges and waste
  - — products and packaging
  - — nuisance and noise
  - — trees, amenities, landscape and wildlife
- Details of:
  - — licences
  - — discharge consents
  - — permits
  - — authorisations

COSHH (Control of Substances Hazardous to Health) Register: health and safety arrangements on the site and the control of substances hazardous to health.

Reports of incidents: failures to comply with policy and remedial actions taken.

Step 7: Scoping study—perform an environmental scoping study, set objectives and targets

Environmental objectives aimed at the continuous improvement of environmental performance should be set at the highest appropriate management level, specified at all relevant levels within the company. Improvements should be quantified wherever possible and the necessary actions given timescales (1).The EC Eco-audit Regulation (1) requires that the site's environmental auditing programme should define in writing the objectives of each audit or audit cycle, including the audit frequency for each activity. These objectives must include, in particular, assessing the management systems in place, and determining conformity with company policies and the site programme, which must include compliance with relevant environmental regulatory requirements.

BS 7750 requires that the organisation should establish and maintain procedures to specify its environmental objectives, and consequent quantified targets consistent with the environmental policy at all relevant levels within the organisation.

The EC Eco-audit Regulation (1) requires that the company should establish and maintain an environmental programme for achieving the environmental objectives at a site. This should include designations of responsibility and the means to achieve these objectives.

The purpose of the **environmental scoping study** is to set the **objectives and targets, plans and strategies** for the forthcoming environmental management exercise and to plan for the **environmental review**. (*What should the environmental review cover? What activities? Site and boundaries? Over what time-scale?*) The overall scope must be clearly defined and should specify the subject areas and activities to be covered, the environmental standards to be considered and the periods covered by the audit/audit cycle.

An action plan and activity schedule must be constructed. The information needed for the review must be identified, as well as particular, possibly sensitive, areas upon which to focus.

Issues to be covered in the environmental programme include (1) all those listed in the **environmental issues**. Information is required to identify and quantify these issues.

Using the information gathered in the initial review:

- Assess the relative importance of environmental impacts
- Decide on the significant issues that need to be covered in the environmental management programme
- Evaluate compliance with current and possible future national and international legislation
- Identify areas for priority action
- Identify training needs
- List data requirements
- Select measurement methods
- Construct information dissemination systems

Step 8: Management—set up/revise environmental management system

The EC Eco-audit Regulation (1) requires that the company's environmental management system should be designed, implemented and maintained in such a way as to fulfil the requirements of:

- environmental policy, objectives and programmes

- defining the organisation, responsibilities and authorities of personnel, including the activity leader, having responsibility of ensuring that the management system is implemented and maintained
- environmental awareness of personnel at all levels, with respect to:
  - the importance of adherence to environmental policy and the achievement of the objectives set
  - the potential environmental effects of their work activities
  - their roles and responsibilities
  - the potential consequences of non-adherence to agreed procedures
  - identifying training needs and providing training
  - documenting the activities

The requirements of BS 7750 with respect to the environmental management system are similar.

If a company has already adopted BS 5750—Quality Systems (6), the requirements of BS 7750—Environmental Management Systems might be latched on to the existing management structure.

Step 9: Documentation—establish/revise documentation systems

This step defines the documentation to be maintained and the registers to be established. A **statement of company commitment** and a **policy statement** on the environment should be produced.

Recording and documentation systems must be established and maintained by the environmental management team. The environmental significances of processes and substances must be identified and a **register of environmental effects** (local and global) should be produced.

This register (2) should adopt a life cycle, i.e. cradle-to-grave, approach and should contain lists of:

- stocks
- consumption of energy
- controlled and uncontrolled emissions to air and their environmental impacts
- consumption of water
- controlled and uncontrolled discharges to water and their environmental impacts
- consumption of materials, solid wastes, waste disposal and their environmental impacts
- hazardous substances and their environmental impacts
- contamination of land
- the use of land, water, fuels and energy and other natural resources
- noise, odour, dust, vibration and visual impact

- effects on specific parts of the environment and ecosystems
- products—inventory, life cycle assessments and environmental impact

Documentation

- Register of regulations
- Environmental review report
- COSHH (Control of Substances Hazardous to Health) Register
- Records of environmental audits
- Reports of incidents, failures to comply with policy and remedial actions taken
- Records of training

Step 10: Review—conduct environmental review

The EC Eco-audit Regulation (1) requires that the company should conduct **environmental reviews** of the site.

The environmental review is a *baseline environmental account* embracing energy, water, materials and waste. The environmental review is a snapshot of an existing plant's environmental performance at one point in time and provides an initial assessment of environmental performance upon which to plan for improvement (5). BS 7750 calls this review the **environmental effects inventory**. The review helps to formulate company objectives and environmental policy. It includes the following components:

Health and safety

- Investigate health and safety arrangements on the site—control of substances hazardous to health (COSHH register)
- Investigate systems for dealing with accidents and emergencies—internal, external—local community, press and media

Legislation and regulations

- Assess compliance with legislation
- Identify and list scheduled processes which are covered by HMIP
- Identify and list scheduled processes which are covered by Local Authority controls
- Classify scheduled processes into those which are covered in Part A and Part B of the IPC regulations of the Environmental Protection Act

Issues

Issues include all those listed in the **ENVIRONMENTAL ISSUES** (cf section 4.2):

- general issues
- processes
- substances
- services
- transportation
- energy
- emissions to air
- water
- discharges to water
- materials
- waste
- other nuisances
- nature and ecology
- accidents, health and safety
- products
- packaging
- information

All associated energy, water, materials flows, emissions to air and water, wastes and discharges, as well as noise and vibrations, odours, etc. should be evaluated.

Assessment, control and reduction of all environmental impacts of the activity should be conducted with respect to:

- energy management and savings
- reduction of gaseous emissions
- water management and savings
- reduction of effluents
- materials management and savings
- reduction of solid waste quantities
- waste avoidance, recycling, reuse, transportation and disposal
- evaluation, control and reduction of noise pollution
- selection of new processes and alterations to existing processes
- product planning (design, packaging, transportation, use and disposal)
- environmental performance and practices of contractors, subcontractors and suppliers
- prevention and limitation of accidents
- accident and emergency procedures
- staff information and training
- provision of external information

Environmental accounting

Environmental accounting is an **input/output** analysis of all flows of energy and materials through a system. It seeks to quantify all inputs, throughputs and outputs to, within and from a system boundary in terms of past, present and future environmental impacts.

An environmental account is a balance sheet of energy and materials *inputs*, *outputs* and *throughputs* concerning the system under scrutiny. Figure 4.2 attempts to summarise the environmental streams flowing through a system.

The system could comprise a built facility, such as a building, a production process or an institution, a product, such as a motor vehicle or a tube of toothpaste, or a service, such as a local council, an educational system, a motorway, or a railway network.

The system boundary could encompass an individual, a home, a family, a place of work, the remit of a local council, a country or the entire world.

Inputs

These include all imports of fossil fuels (coal, oil (petrol, diesel lubricating oil, etc.), natural gas, liquid petroleum and other gaseous fuels), electricity, raw materials, water, air for combustion or respiration, metals and alloys, and non-metals (mineral matter, chemicals and plastics, process fluids and gases, organic matter, including food, paper, livestock, people) and manufactured goods.

Referring to Figure 4.2, the energy inputs to the system comprise solar energy, fossil fuels and electricity, people and livestock. The system also needs air and water.

The *inputs* are in general:

- materials supplied, including fossil fuels and raw materials and their historical energy and environmental contents
- electricity supplied and its historical energy and environmental content
- processed material supplied, including water and its historical energy and environmental content
- products supplied and their historical energy, materials and environmental content
- people and livestock supplied and their energy and environmental requirements attributable to the system under scrutiny
- the energy and environmental contents of recycled materials which are rejected, processed outside the system and returned to the system
- solar energy and other environmental gains
- pollution imported, such as waste effluent, polluted gases, acid rain and solid wastes

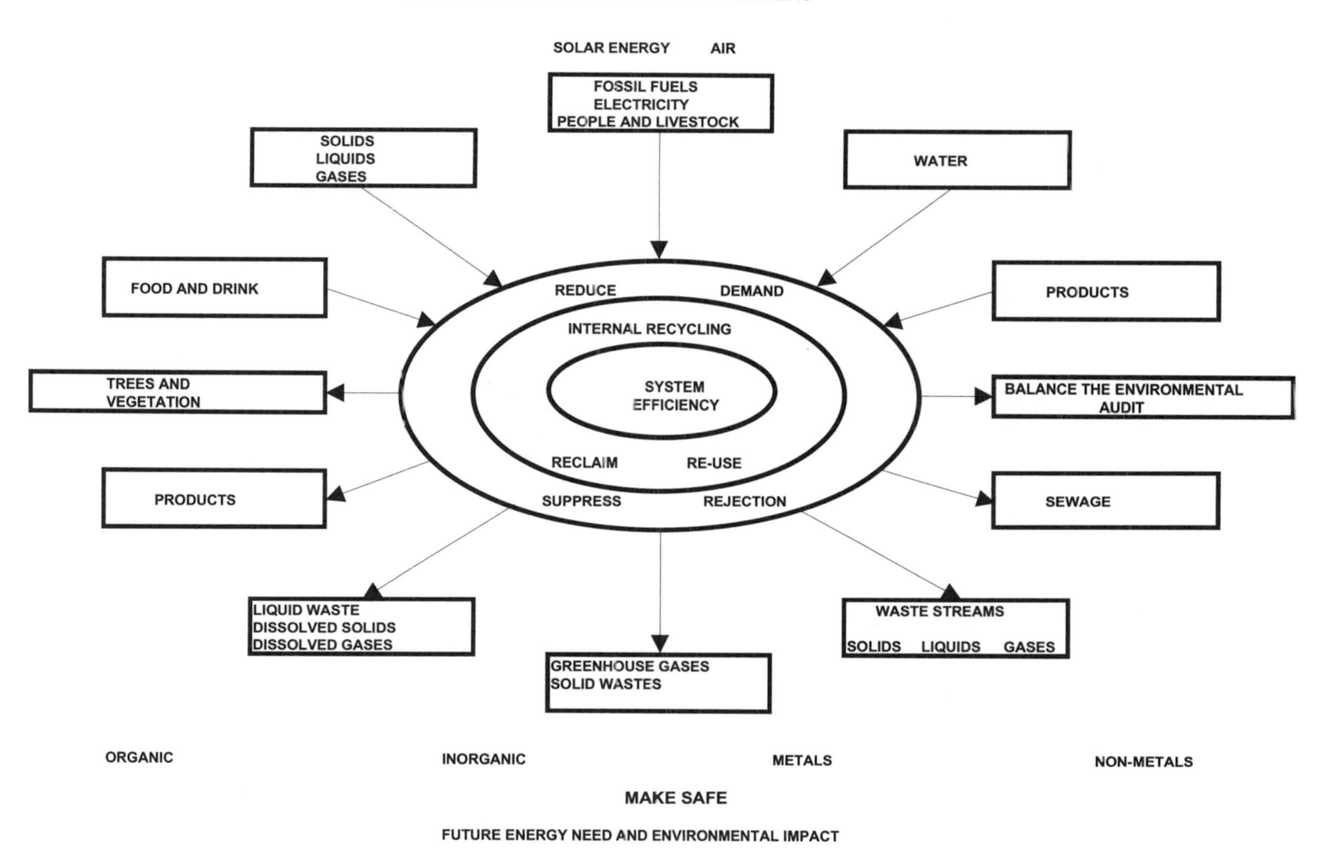

Figure 4.2 Environmental streams flowing through a system

Past environmental impacts?

The item imported will have a history of energy and materials absorbed and environmental interaction. It will have acquired a historical input of energy and materials and produced waste matter. These must be quantified. The fuels and materials will have been mined, extracted, processed and transported. The electricity would have been produced from the combustion of fossil fuels, producing waste gases and thermal pollution of air and water, or by nuclear reaction, producing radioactive wastes and thermal pollution of water. The water supplied will have been collected, purified and pumped to its point of use. The food will have been grown, producing land and water pollution, harvested using fossil fuels for farm machinery, transported, processed, distributed and sold, requiring energy and materials and producing pollution at every stage of this chain. The livestock will have required food, land, shelter and transportation. The manufactured goods will be constructed from all these building blocks and the people will be consumers of all these commodities.

Throughputs

The *throughputs* are all the inputs which are converted to outputs, taking account of:

- energy and materials which are internally recycled
- plant matter which has undergone photosynthesis and is used internally

For each throughput item, its historical energy content must be estimated in terms of quantities of primary fuels required and energy released (MJ). This may be converted to the historical air pollution associated with the combustion processes involved. If the throughput item is combusted, the output air pollution may be calculated.

Outputs

The fossil fuels leave the system as greenhouse gases, smoke and solid wastes. Water leaves as liquid waste containing dissolved solids, liquids and gases. Water, food and drink leave the system as sewage. Other solids, liquids and gases may be processed by the system and rejected as pollutants from the system.

The *outputs* from the system are in general:

- gaseous, liquid and solid materials and their future environmental impacts
- fuels and electricity
- products and their future environmental impacts
- people and livestock exported and their energy and environmental requirements attributable to the system under scrutiny

Present environmental impacts?

The direct combustion of fossil fuels and the release of gaseous, liquid and solid waste to the system environment become direct environmental impacts from within the system. Quantities of waste items may be calculated. Some of these may become direct outputs by dispersal across the system boundary, such as smoke and waste gases dumped to the atmosphere and carried away by the winds, or waste liquid effluents dumped to watercourses, rivers or seas.

Future environmental impacts?

Solid wastes, liquids and gases output from the system boundary need to be processed using energy and materials, combusted or disposed outside the system. The resulting future pollution must be calculated to assess the future environmental impacts.

Environmental accounting seeks to assign numerical values to the quantities and qualities of the input, throughput and output environmental streams so that the system's degree of environmental imbalance may be quantified. This imbalance may be reduced by

- planting trees and vegetation to absorb carbon dioxide
- reducing the system's demands for energy, water, other materials and products by good energy, water and materials management
- reducing rates of rejection of heat, water, materials and products
- recycling energy, water and materials
- 'locking up' unavoidable pollutants

An **energy account** may then be constructed by summing the historical and throughput energy requirements for all components of the system. This may be converted to the **energy-associated environmental accounts** by calculating the concomitant air pollution.

The third stage is to conduct a similar input/throughput/output analysis of non-energy-associated materials, such as water, solid and gaseous wastes, to produce the **non-energy-associated environmental accounts**.

Tables 4.5 to 4.7 summarise the three environmental accounts which need to be constructed for any system for a representative period (usually over one representative year).

Balancing the environmental accounts

Since energy and materials are conserved, the above accounts will already be in balance. However, high grade feedstock, energy and materials have been down-graded to environmental pollutants, which must be made safe. The balanced

Table 4.5 The energy accounts (MJ, £)

The energy accounts	**Quantities** MJ	£
Inputs		
Direct fossil fuels and electricity		
Coal		
Oil-derived fuels		
Gas		
Other fuels		
Electricity (primary energy content)		
Historical energy contents of:		
imported materials		
imported products		
people		
Ouputs		
Energy in products		
Energy rejected to the environment		
Energy in waste products		

Table 4.6 The energy-related environmental accounts (kg CO_2 equivalent, SO_2 etc.)

The energy-related environmental accounts	**Quantities** kg CO_2 equivalent	kg SO_2 etc.
Inputs		
Direct fossil fuels and electricity		
Coal		
Oil-derived fuels		
Other fuels		
Electricity (primary environmental content)		
Historical environmental content of:		
imported materials		
imported products		
people		
Locked in system		
Trees, plants, etc.		
Outputs		
Locked in products		
Intrinsic in waste products		
Rejected to the environment		

Table 4.7 The non-energy-related environmental accounts (kg waste material × environmental toxicity rating)

The non-energy related environmental accounts	Quantities kg	£
Inputs		
Water		
Air		
Outputs		
Polluted air (non-energy-related)		
Liquid effluents		
Solid wastes		

Note
The environmental toxicity rating should consider estimates of the future energy and materials required to neutralise or recycle the material or for safe long-term storage.

environmental account produces no net CO_2 equivalent or acid or toxic gases, liquid effluent or solid wastes.

The system's degree of imbalance may be improved by:

- reducing demands for energy, materials and products by energy and materials (including water and waste) management and recycling—*this option saves energy and materials, saves money and protects the environment*
- reducing the emissions of gases by flue gas clean-up and waste gas management—*this option protects the environment but costs money*; it would reduce pollution taxes, if levied
- increasing the internal carbon fixers, e.g. by planting trees—*this option enhances and protects the environment but costs money*; it should reduce pollution taxes, if levied

The aim of environmental balancing is to:

Save energy and materials
Save money
Protect the environment

In summary, the environmental accounts must include:

For the site:

- Input/throughput/output analyses and accounts of the utilisation of energy, water and materials—quantities and qualities
- Measurements of energy and materials usage, emissions, effluents and waste discharges

For transportation:

- Quantification of energy consumed, noise and pollution produced in transportation to, within and from the site
- An environmentally sound transport policy—routing and timing

For products, packaging and materials:

- Product and process design, planning and management
- Evaluation of packaging requirements—quantities and materials used, recycleability

Systematic environmental accounting thus constructs a quantitative input/throughput/output analysis across a site or within a product life cycle.

Chapters 5, 6 and 7 deal with energy, water, materials and waste accounting in further detail.

Seven steps to conducting the environmental review

(1) Identify all activities, inputs and outputs.
(2) Characterise and quantify all environmental burdens.
(3) Identify environmental effects.
(4) Determine significant environmental effects.
(5) Identify environmental requirements.
(6) Determine status of compliance with legislative requirements.
(7) Review environmental management arrangements.

Documentation

The environmental review report and register of environmental effects (2), which should contain lists of:

- controlled and uncontrolled emissions to the atmosphere
- controlled and uncontrolled discharges to water
- solid and other wastes
- contamination of land
- use of land, water, fuels and energy and other natural resources
- noise, odour, dust, vibration and visual impact
- effects on specific parts of the environment and ecosystems

This should adopt a life cycle, i.e. *cradle-to-grave* approach.

Step 11: Options/recommendations—identify and evaluate options for improvements and make recommendations

Environmental improvements

Having constructed the environmental accounts, it is possible to identify where potential cost savings may be made in the areas of energy, water and materials utilisation and waste disposal. For example:

- Are these essential?
- Are there alternatives which are more environmentally acceptable?
- Can consumption rates be reduced, minimised or eliminated?
- Are there recycling possibilities?

Chapters 5, 6 and 7 deal with energy management, water management and waste management, identify options for improvement and provide checklists for environmental improvements. These options may:

- save money by reducing fuel consumption
- save money by reducing electricity consumption
- save money by reducing water consumption
- save money by reducing materials consumption
- save money by minimising waste
- save money by reducing effluent charges resulting from effluent reduction
- save money by reducing solid waste disposal charges associated with the reduction of materials utilisation and amounts of solid wastes
- save money by reducing carbon taxes resulting from lower amounts of carbon emissions resulting from greater energy use efficiencies
- save money by avouding the necessity of paying VAT on fuel wastage or inefficiencies
- save money by avoiding the necessity of paying a pollution tax
- save money by avoiding fines for pollution offences
- save money by avoiding the necessity of paying for remediation of environmental pollution damage
- save money by avoiding the necessity of paying financial compensation for losses endured by third parties
- save loss of peace of mind by avoiding imprisonment
- make money by achieving better public relations
- make money by increasing competitive advantage
- save money by obtaining cheaper loans
- save money by obtaining cheaper insurance cover
- make money by attracting shareholders
- make money by retaining and gaining customers for the company's products and services

- continue to make money by satisfying suppliers and trading partners that the company is conforming to recognised environmental standards
- make money by attracting, retaining and motivating a quality workforce

There is no doubt that the costs of energy, electricity and water supplies and raw materials will rise substantially in real terms, as the supply companies comply with new environmental legislation and standards. Pollution taxes, effluent charges and waste disposal costs will also rise as the effluent and waste disposal operators also have to comply with high environmental standards and legislation.

Financial returns

Options for financial investments and returns may then be identified and evaluated to construct an *optimal investment schedule*, as well as a *no-risk investment plan* for the future, where perceived and realised financial 'savings' are reinvested to allow subsequent more cost-effective investments to be made. The construction of such an investment schedule is demonstrated in Chapter 8.

The systematic application of integrated environmental management therefore allows all necessary retrofits and modifications to technical activities to be financially resourced from within the environmental management project itself.

Step 12: Audit/report—conduct environmental audit and report, conduct periodic environmental audits

The company must carry out periodic environmental audits of the sites concerned. The European Commission defines environmental auditing as:

> A management tool comprising a systematic, documented, periodic and objective valuation of how well organisations, management and equipment are performing with the aim of contributing to safeguard the environment by:
>
> - facilitating management control of environmental policies
> - assessing compliance with company policies, which would require meeting regulatory requirements

Environmental auditing is a relatively new term which is still being defined and revised by the European Commission, the Confederation of British Industries and other bodies. The British 'Association of Environmental Consultancies' has proposed the following definition:

> A full environmental audit comprises a complete examination of all aspects of a company's interactions with the environment. It is designed to examine performances in terms of:
>
> - Management
> - Compliance with policies and regulations
> - Information to be disclosed to the authorities or to the public

The full environmental auditing procedure as proposed by the European Commission is outlined below. The audit should include at least the following steps, which should be documented:

- planning of the audit activities, including a definition of responsibilities
- review of the environmental protection policies of the company
- assessment of the organisation, its management, and its equipment
- gathering of data and all relevant information
- evaluation of overall performance
- identification of areas for improvement
- internal reporting to top management

A follow-up policy statement by top management should conclude the auditing procedure.

At least the following aspects should be considered by the audit:

- relevant information and data on material and energy exchanges with the environment due to the activity being audited
- compliance with community, national and local regulations and standards applicable to the establishments concerned
- significant incidents or operation disturbances occurred, or complaints received

There should also be a review of the available policy, equipment, organisation and management systems related to the following aspects:

- pollution discharge control, monitoring and reduction
- other nuisances
- energy choice and reduction of energy use
- raw materials choice and transportation—reductions of water and raw materials uses
- waste transportation, elimination, recycling and reuse
- product planning (design, packaging, transportation, use and elimination)
- prevention and mitigation of accidents
- staff training and participation
- provision of external information

Internal environmental audit report

This should contain information as follows:

(1) *Executive summary*
    - Strengths of the organisation
    - Weaknesses of the organisation
    - Opportunities for the organisation
    - Threats to the organisation

- Environmental, health, safety, legal and financial risks raising areas for concern
- Legal considerations
- Licences needed
- Adherence to standards
- *Overall summary of recommendations*

(2) *Site description*

- Land system boundaries
- Inventory of land involved
- History
- Historical review of the site
- Historical uses for land
- Environmental hazards from past uses
- Contaminated ground or groundwater
- Environmental impacts
- Current uses of land
- Present operations
- Maintenance programmes
- *Scope*
- *Aims of the work*

(3) *Processes*

- Selection of new processes and alterations to existing processes
- Assessing impacts of new activities in advance
- Minimising the environmental effects of new activities by advance planning
- Functions of processes
- Compliance with legislation and regulations
- Scheduled processes covered by HMIP
- Scheduled processes covered by Local Authority controls
- Scheduled processes covered in Part A of the IPC regulations of the Environmental Protection Act
- Scheduled processes covered in Part B of the IPC regulations of the Environmental Protection Act
- Impacts on the local environment
- Impacts on the global environment
- Minimising waste of energy, water and materials by process improvement
- Measures necessary to minimise, prevent or eliminate pollution
- Monitoring
- Measurements
- Testing
- Controls
- *Summary of recommendations*

(4) *Substances*
- Types
- Inventory
- Sources
- Quantities
- Quality
- Uses
- Frequencies of use
- Costs
- Stocks
- Hazardous substances and environmental impact
- Monitoring
- Controls
- Transport
- *Summary of recommendations*

(5) *Services*
- Buildings inventory
- Facilities inventory
- Current usage
- Past usage
- Current environmental effects of past usage
- Present environmental hazards (e.g. asbestos insulation)
- Building services
- *Summary of recommendations*

(6) *Transportation*
- Traffic pollution produced in transportation to, within and from the site
- Energy consumed in transportation to, within and from the site
- Noise produced in transportation to, within and from the site
- Air pollution produced in transportation to, within and from the site
- Formulation of an environmentally sound transport policy—vehicles, fuels, routes and time schedules
- *Summary of recommendations*

(7) *Energy* and *emissions to air*
Energy
- Energy choice and types
- Sources
- Quantities
- Qualities
- Consumption figures
- Fuels and electricity costs
- Stocks

- Fuels and electricity delivered
- Uses of energy (fuels and electricity)
- Energy management
- Energy rejection to the environment
- Conservation and energy savings
- *Summary of recommendations*

Emissions to air

- Complaints received
- Controlled and uncontrolled emissions to air
- Energy-related emissions
- *Summary of recommendations*

(8) *Other emissions to air*

- Types
- Substances
- Sources
- Quantities
- Qualities
- Frequencies of emissions
- Dispersion
- Monitoring
- Environmental impact
- Complaints received
- Needs for clean-up
- Reduction of emissions
- Discharge control
- Rendering harmless unavoidable emissions
- *Summary of recommendations*

(9) *Water* and *discharges to water*

Water

- Water delivered
- Water consumption
- Costs for water
- Uses for water
- Water flows
- Water wastage
- Water management
- *Summary of recommendations*

Discharges to water

- Complaints received
- Inventory of discharges to water
- Inventory of effluents
- Energy-related discharges to water

- Other controlled and uncontrolled discharges of wastes to water
- Controlled and uncontrolled effluents
- Sinks for wastes to water and effluents
- Liquid wastes (hazardous or other) and environmental impact
- Discharge monitoring
- Water sampling
- Discharge control
- Reduction of effluents
- Effluent clean-up
- Needs for water clean-up
- *Summary of recommendations*

(10) *Materials* and *waste*

Materials

- Raw materials choice
- Stocks
- Storage arrangements
- Environmental impact of raw material sourcing
- Controlling the environmental effects of raw material sourcing
- Inventory and materials accounts
- Materials delivered
- Uses of materials
- Materials flows
- Materials management
- Transportation of materials
- Toxic materials
- Radioactive materials
- Environmental impact
- *Summary of recommendations*

Waste

- Inventory of wastes
- Sources of waste
- Analyses of waste
- Sinks for waste
- On-site waste disposal methods
- Contamination of land
- Environmental impacts
- Complaints received
- On-site waste management
- Waste disposal
- *Summary of recommendations*

(11) *Other nuisances*

- Thermal energy

- Noise
- Vibration
- Dust
- Odours
- Monitoring
- Controls
- *Summary of recommendations*

(12) *Nature* and *ecology*
- Effects on ecosystems
- Nature conservation
- Litter
- Visual impact
- Landscaping
- Trees
- Plants
- Wildlife
- Protecting local ecology
- Protecting global ecology
- *Summary of recommendations*

(13) *Accidents, health and safety*
- Prevention and mitigation of accidents
- Measures necessary to prevent accidental emissions of energy or materials
- Health and safety considerations and arrangements on the site
- Control of special substances hazardous to health (COSHH register) and prescribed substances
- Procedures for handling special and prescribed substances
- Spill control measures
- Fire control measures
- Systems and procedures for dealing with accidents and emergencies
- Monitoring
- Controls
- Systems for dealing with the local community, press and media
- *Summary of recommendations*

(14) *Products*
- Inventory of imported products
- Uses of imported products
- Reduction, minimisation or elimination of imported products
- *Summary of recommendations*
- Inventory of *exported* products
- Product planning
- Product life cycles—cradle-to-grave environmental impact assessments

- Designing products to minimise their overall cradle-to-grave environmental impact
- Waste avoidance in product manufacture
- Product transportation
- Uses of exported products
- Energy consumption of exported products
- Water consumption of exported products
- Materials consumption of exported products
- Air pollution by exported products
- Water pollution by exported products
- Waste production by exported products
- Reclamation of products
- Reuse of products
- Recycling of products
- Ultimate disposal of products
- *Summary of recommendations*

(15) *Packaging*

- Paper and packaging policies
- Packaging of *imports*
- Inventory of imported packaging
- Packaging requirements—materials, quantities and qualities used
- Use of imported packaging
- Need for imported packaging
- Disposal of imported packaging
- Recycleability of imported packaging
- Recycling of imported packaging
- *Summary of recommendations*
- Packaging of *exports*
- Inventory of exported packaging of products
- Packaging requirements—materials, quantities and qualities used
- Transportation of exported packaging
- Reclamation of exported packaging
- Reuse of exported packaging
- Recycleability of packaging
- Recycling of exported packaging
- Ultimate disposal of product packaging
- *Summary of recommendations*

(16) *Legal considerations and adherence to standards*

- Prescribed substances
- Prescribed processes
- Problems arising
- Incidents and reactions

- BATNEEC
- BPEO
- Licences needed
- *Summary of recommendations*

(17) *Information*
- Education
- Staff training
- Skills
- Records
- Registers
- Internal information
- Staff participation
- External information
- Environmental performance and practices of contractors, subcontractors and suppliers
- *Summary of recommendations*

(18) *Conclusions and recommendations*
- Summary of achievements
- Recommendations, costings and expected returns (accuracies of estimates)
- Recommendations for further work
- *Overall summary of recommendations*

4.6 VALIDATION OF THE ENVIRONMENTAL MANAGEMENT SYSTEM

Environmental management audits usually review environmental management systems, focusing on the development of policy, appropriate procedures and systems, and product life cycle (cradle-to-grave) analytical techniques.

Independent validation

The EC Eco-audit Regulation (1) requires that a company environmental management system should include environmental auditing procedures to help management assess compliance with the system and the effectiveness of the system in fulfilling the company's environmental policy. It differs from the requirements of BS 7750 (i.e. that the organisation identify in-house verification requirements and procedures) in that it states that, in order for a company to gain EC Eco-audit accreditation, it is necessary for it to provide for independent and neutral accreditation and supervision of environmental verifiers in order to ensure the credibility of the environmental management scheme, that is:

- environmental policy
- environmental programme
- environmental management system
- environmental review procedure
- environmental audit procedure
- environmental statement

Thus, whereas the auditor can be an individual or a team belonging to the company or external to the company, acting on behalf of top management, the accredited environmental verifier must be a person or organisation, independent of the company being verified, who has obtained accreditation within the member states of the EU.

On-site audit activities by the external verifier will include discussions with site personnel, inspections of operating conditions and equipment and reviewing of records, written procedures and other relevant documentation, with the objective of evaluating environmental performance at the site by determining whether the site meets the applicable standards and whether the system in place to manage environmental responsibility is effective and appropriate.

The following steps must be included in the validation process:

- understanding the management system
- assessing the strengths and weaknesses of the environmental management system
- gathering relevant evidence
- evaluating audit findings
- preparing audit conclusions
- reporting audit findings and conclusions to top company management

This report should include:

- the scope of the external audit
- the state of compliance with the company's environmental policy
- the environmental progress on the site
- the effectiveness and reliability of the arrangements for monitoring environmental impacts at the site
- demonstrations of the need for corrective action, where appropriate
- the preparation of a plan of appropriate action
- methods for ensuring the implementation of this action plan to ensure that the audit results are followed up

The EC Eco-audit Regulation (1) suggests that the audit should be completed, as appropriate, at intervals not exceeding three years, depending upon the nature, scale and complexity of the activities, emissions, waste, raw materials, water and energy consumption, and the general interaction with the environment, as well as the importance, urgency and histories of the problems detected at the site.

BS 7750 (2) requires that the entire environmental management system be reviewed periodically by internal independent verifiers.

Documentation

- Records and reports of the internal environmental audits
- Reports of the audit by the external validator
- Periodic reviews of the environmental management system

4.7 THE 39 SEQUENTIAL STEPS OF INTEGRATED ENVIRONMENTAL MANAGEMENT

1. Identify education and training needs.
2. Set up training programmes.
3. Make a statement of company commitment.
4. List environmental issues.
5. Conduct an initial review.
6. Set company policy.
7. Set company objectives.
8. Set company targets.
9. Construct a company policy statement.
10. Make an environmental statement.
11. Structure the environmental organisation.
12. Construct a register of regulations.
13. Assess compliance with legislation.
14. Perform an environmental compliance audit.
15. Carry out scoping study.
16. Determine the objectives of the work to be done.
17. Set the targets for the work to be done.
18. Construct an environmental management programme.
19. List the action plan.
20. Construct an activity schedule.
21. Identify information requirements.
22. Set up an environmental management system.
23. Document the register of environmental effects.
24. Compile an environmental management manual.
25. Perform an environmental review.
26. Construct environmental accounts.
27. Report on the environmental review.

28. Identify and evaluate improvement options.
29. Produce a prioritised investment portfolio.
30. Construct an optimal investment schedule/project plan.
31. Set the target cash flow forecast.
32. Monitor the project.
33. Audit the environmental management system.
34. Make an internal environmental management audit report to top management.
35. Internal reportage and dissemination.
36. External reportage and dissemination.
37. External validation of the environmental management system.
38. Review the entire system and process.
39. Go back to step 1 and start over. . .

REFERENCES

(1) EC Council Regulation allowing voluntary participation by companies in the industrial sector in a Community eco-management and audit scheme, Council Regulation (EEC) No. 1836/93, *Official Journal of the European Communities*, No. L 168/1, 10 July 93.
(2) *BS 7750, Specification for Environmental Management Systems*, British Standards Institution, 1992.
(3) A. Porteous, *Dictionary of Environmental Science and Technology*, Open University Press, Buckingham, 1991.
(4) K. Sadgrove, *The Green Manager's Handbook*, Gower Press, Aldershot, 1992.
(5) R. Welford and A. Gouldson, *Environmental Management and Business Strategy*, Pitman Publishing, London, 1993.
(6) *BS 5750, Specification for design, development, production, installation and servicing Quality Systems*, British Standards Institution, 1987.

5

Energy, emissions to air and energy management

This chapter deals with fossil fuel combustion and the concomitant polluting emissions to air. The types and calorific values of the various fuels are reviewed and example calculations of energy release and pollutants produced are provided. Both energy-related and non-energy-related emissions to air are considered. Rules for the efficient and cost-effective conservation of energy are included, as well as questions for the client interview and a checklist for energy managers. Thirty-nine steps to optimise energy management investments at a site are constructed. The procedures of energy management are demonstrated via a worked example. It is shown that a site can simultaneously save energy, save money, reduce pollution and so protect the environment.

5.1 FOSSIL FUELS

Fossil fuels consist of the combustible elements: hydrogen, H_2, carbon, C, and sulphur, S. They often also contain the incombustible element nitrogen, N_2, moisture, H_2O, and minerals.

When combusted, the carbon oxidises to carbon monoxide, CO, (if insufficient air is present) and carbon dioxide, CO_2, the hydrogen oxidises to water (steam), H_2O, and the sulphur (in coals and liquid fuels) becomes sulphur dioxide, SO_2.

The carbon and sulphur dioxides dissolve in water to form carbonic acid, H_2CO_3, and sulphuric acid, H_2SO_4, which contribute to acid rain.

When combustion temperatures are high, nitrogen in air can combine with oxygen to form the NOx gases, e.g. nitrous oxide, N_2O, which dissolved in water forms nitric acid, HNO_3. The action of sunlight on exhaust fumes can also form nitrous oxides and ozone, O_3, the main contributors to photochemical smog.

Moisture present in a fuel absorbs sensible heat from the flame to supply the latent heat needed to change phase from liquid to vapour. Any minerals present form the residual ash.

Smoke is a dispersion in air of small solid particles of uncombusted fuels (carbon and sulphur) with some airborne ash.

Volatile organic compounds (VOCs) are non-combusted organic liquids, such as petrol and benzene.

End uses for fossil fuels

In the UK, fossil fuels are used in power generation, industry, domestic, commercial and public administration (DCP) sectors, agriculture and transport. Figures 5.1a and b show that the total rate of energy consumption rose by 17% over the period from 1960 to 1994, this being almost entirely due to the rise in energy consumption by transport (a 238% rise over the period, increasing at 4.2% per year), whilst consumption by the other sectors remained stable or declined.

Coal

Fossilised reserves of coal emanate from land-based vegetable matter (wood converting to peat and then to coal over time). Heating coal in the absence of air yields four main products: coal gas, ammoniacal liquor, coal tar and coke.

Typical compositions of some of the solid fuels are listed in Table 5.1.

Table 5.1 Typical compositions of solid fuels

Substance	% C	% H_2	% S	% N_2	% O_2	% H_2O	% Ash
Wood	46.0	4.0	0.1	0.5	37.5	9.3	3.0
Lignite	52.2	4.0	0.9	0.5	17.0	18.4	7.2
Coal-4000	44.0	3.1	0.3	0.8	4.8	3.8	43.0
Coal-4500	49.0	3.8	0.5	1.0	5.4	6.0	34.4
Coal-5500	59.0	3.2	0.4	1.0	4.9	6.0	25.2
Husk	36.0	3.7	0.1	0.5	29.3	9.0	19.4
Bagasse	41.0	4.6	0.04	0.2	36.7	10.5	7.0
Coke	90.0	1.6	1.0	1.0	1.4	0.0	5.0

Note
See also values given in (2).

Oil

Both oil and gas reserves emanate from marine creatures (plankton reducing to methane, a major component of natural gas, and then forming oil over time). Petroleum oils consist of hydrocarbons (mainly alkanes), aromatic hydrocarbons and some sulphur, oxygen and nitrogen containing compounds.

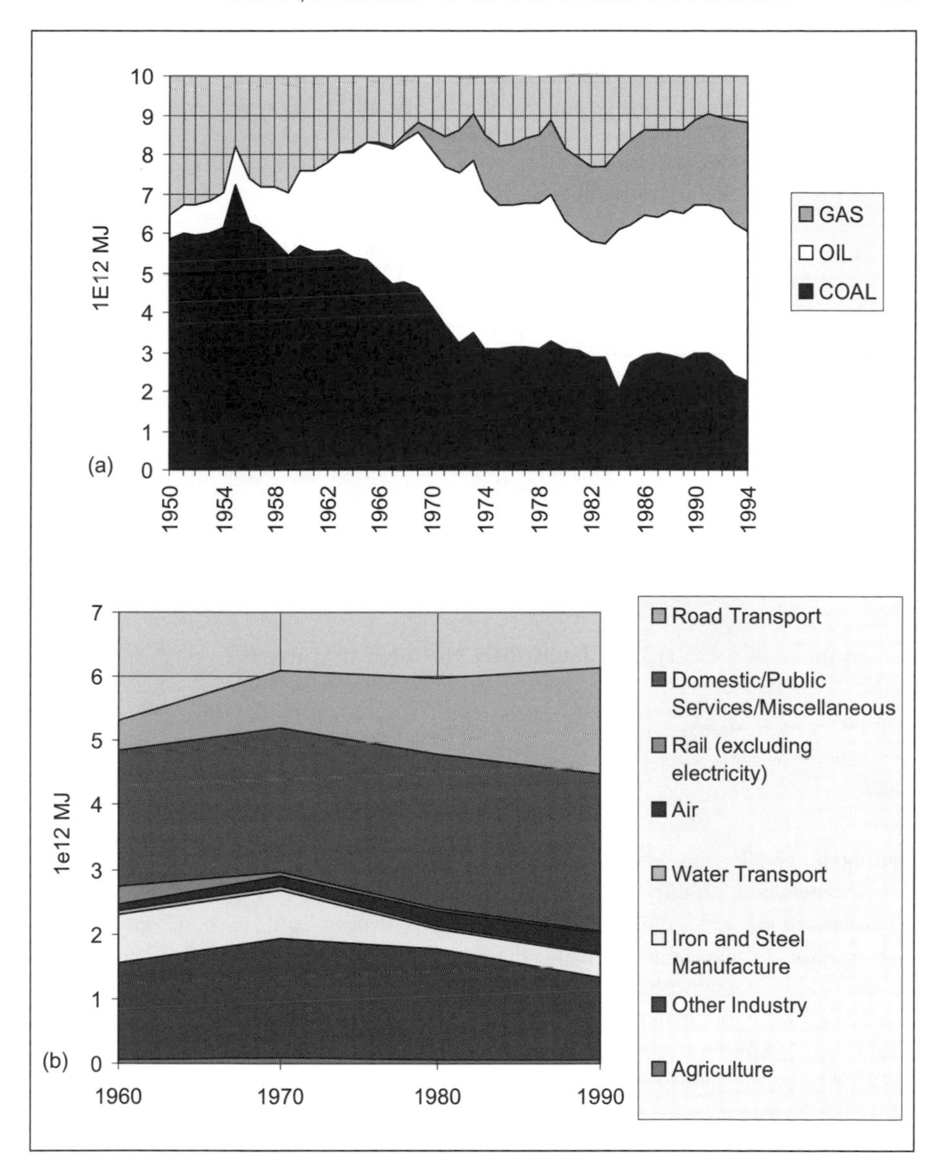

Figure 5.1 (a) UK fossil fuel consumption; (b) UK end uses for energy delivered to final consumers

Source: Digest of Environmental Protection and Water Statistics, UK Statistical Office, Department of the Environment, 1992.

Table 5.2 Typical compositions of liquid fuels

Fuel	% C	% H	% S
Motor petrol	85.5	14.4	0.1
Vapourising oil	86.8	12.9	0.3
Kerosene	86.3	13.6	0.1
Diesel (gas) oil	86.3	12.8	0.9
Light fuel oil	86.2	12.4	1.4
Heavy fuel oil	86.1	11.8	2.1
Orimulsion	84.0	10.2	4.15

Liquid fuels

- Paraffins, C_nH_{2n+2}
- Olefins and naphthenes, C_nH_{2n}
- Aromatics, C_nH_{2n-6}

Compositions of some typical liquid fuels are listed in Table 5.2.

Gases

Hydrogen, H_2, having no carbon content, has the highest calorific value of the gaseous fuels and produces only water when combusted.

Hydrocarbons are compounds containing hydrogen and carbon only. Most hydrocarbons in petroleum deposits occur naturally as liquids, but a few exist in the gaseous phase at atmospheric temperatures and pressures.

Alkanes form the major constituents of crude petroleum and are also present in natural gas from underground wells. The general formula for this homologous series of saturated hydrocarbons is C_nH_{2n+2}. Samples of alkanes are listed in Table 5.3.

Alkenes are unsaturated hydrocarbons having the general formula C_nH_{2n}. They include the substances listed in Table 5.4.

Alkynes are unsaturated hydrocarbons which have the general formula C_nH_{2n-2}. They include those substances listed in Table 5.5.

Natural gas is comprised mainly of methane, but contains carbon dioxide as well as small percentages of ethane, propane, butane, pentane, hexane, nitrogen and oxygen (2).

A typical ultimate composition for natural gas is given in Table 5.6 and typical compositions of other gaseous fuels are listed in Table 5.7.

Table 5.3 Alkanes

Number of carbon atoms in chain	Alkane	Formula	Boiling point	State at STP
1	Methane	CH_4	−161°C	Gas
2	Ethane	C_2H_6	−89°C	Gas
3	Propane	C_3H_8	−42°C	Gas
4	Butane	C_4H_{10}	−0.5°C	Gas
5	Pentane	C_5H_{12}	36.1°C	Liquid
6	Hexane	C_6H_{14}	69.0°C	Liquid
7	Heptane	C_7H_{16}	98.4°C	Liquid
8	Octane	C_8H_{18}	18 compounds	Liquid
9	Nonane	C_9H_{20}		
10	Decane	$C_{10}H_{22}$		

Note
STP = standard temperature and pressure, 25°C and 1 atmosphere.

Table 5.4 Alkenes

Number of carbon atoms in chain	Alkene	Formula	Boiling point	State at STP
2	Ethene, ethylene	C_2H_4	−105°C	Gas
3	Propene, propylene	C_3H_6	−48°C	Gas

Table 5.5 Alkynes

Number of carbon atoms in chain	Alkyne	Formula	Boiling point	State at STP
2	Ethyne, acetylene	C_2H_2	−83.4°C	Gas
3	Propyne	C_3H_4	−23.3°C	Gas

Table 5.6 Typical composition of natural gas

Natural gas	% C	% H_2	% S	% N_2	% O_2	% H_2O	% CO_2
	66.0	23.0	0.0	4.0	0.0	1.5	6.0

Table 5.7 Typical compositions of other gaseous fuels

Gases	H_2	CO	CH_4	C_2H_4	C_4H_8	O_2	N_2	CO_2
Coal gas	49.4	18.0	20.0	0.0	2.0	0.4	6.2	4.0
Producer gas	12.0	29.0	2.6	0.4	0.0	0.0	52.0	4.0
Blast furnace gas	2.0	27.0	0.0	0.0	0.0	0.0	60.0	11.0

5.2 COMBUSTION

Atomic weights

The atomic weights for the constituent atoms of the fossil fuels are given in Table 5.8 and Table 5.9 lists the molecular masses of the elements in fossil fuels.

A **mole** is the molecular mass of a substance expressed in **grams**.

Calorific values

The **calorific value** of a combustible substance is the heat generated when 1 kg of that substance is completely burned (also known as the *heat of combustion* or *heat value* for the fuel).

The **net (or lower) calorific value** for a fuel is calculated when the water, H_2O, in the combustion products is in its vapour form. The **gross (or higher) calorific value** for a fuel is calculated when the water in the combustion products is in its liquid form. The latent heat of vaporisation of water is 2.5 MJ kg^{-1}.

Table 5.8 Atomic weights

H	1
C	12
S	32
O	16
N	14

Table 5.9 Molecular masses

H_2	2
C	12
S	32
CH_4	16
O_2	32
H_2O	18
CO	28
CO_2	44
SO_2	64
N_2O	44

Measurement of calorific values

The calorific value of a fuel is measured in calorimetric tests where the products of combustion are cooled to normal atmospheric conditions. The total heat released in the process of cooling is measured. During this process, water vapour present in the combustion products condenses, releasing its latent heat, and resulting in more heat being liberated than would be available had no condensation occurred. The calorific value indicated by such tests is thus the gross calorific value. The net calorific value is then obtained by subtracting the latent heat of the water present from the gross calorific value.

Table 5.10 lists calorific values for various fuels.

Table 5.10 Calorific values of various fuels (MJ/kg)

Fuel	Formula	Molecular weight	Higher calorific value (MJ/kg)	Lower calorific value (MJ/kg)
Hydrogen gas	H_2	2	143.4	120.9
Liquid hydrogen	H_2	2	94.0	116.5
Methane[1]	CH_4	16	55.83	50.2
Acetylene (ethyne)	C_2H_2	26	50.01	48.28
Ethylene	C_2H_4	28	50.47	47.24
Ethane	C_2H_6	30	52.13	47.63
Propene (propylene)	C_3H_6	42	48.8	45.58
Propane	C_3H_8	44	50.59	46.49
Butane	C_4H_{10}	58	49.5	45.62
Pentane	C_5H_{12}	72	48.9	45.15
Hexane (liquid)	C_6H_{14}	86	48.0	44.33
Toluene (methyl benzene)	C_7H_8	92	47.8	46.0
Heptane	C_7H_{16}	100	48.0	44.44
Octane (liquid)	C_8H_{18}	114	47.6	44.05
n-Decane (liquid)	$C_{10}H_{22}$	142	47.3	43.825
Fuel oil			44.0	41.8
Petrol[2]			44.0	41.8
Crude petroleum, oils			41.8–48.7	38.6–45.5
Kerosene[3]	CH_2	14	43.0	39.8
Residual (heavy) fuel oil			40–45	37–42
Orimulsion			43.0	40.0
Benzene	C_6H_6	78	42.2	40.5
Diesel oil			42.0	38.5
Naphthalene (solid)	$C_{10}H_8$	128	40.1	
Scrap tyres			38.0	
Body fat			37.6	
Bio-oil			34.8	
Biogas			34.0	
Carbon	C	12	32.8	32.8

continued overleaf

Table 5.10 *(continued)*

Fuel	Formula	Molecular weight	Higher calorific value (MJ/kg)	Lower calorific value (MJ/kg)
Town gas			32.0	
Coke			26–35	
Anthracite			29.9	
Bituminous coal			26–32	
Ethanol (ethyl alcohol, liquid)	C_2H_6O	46	29.6	
Coffee grounds			23.2	
Corn cobs			22.0	
Cottonseed cake			22.0	
Municipal refuse			22.0	
Pine bark			22.0	
Wood (dry)			20–21.3	
Wheat straw			19.7	
Bagasse (sugarcane residual)			19.5	
Wood (20% moisture)			16.2	
Lignite			14.6–19.2	
Cattle manure (dry)	cellulose		17.1	
Carbohydrates[4]			16.7	
Sugar (sucrose)	$C_{12}H_{22}O_{11}$	342	16.0	
Glucose	$C_6H_{12}O_6$	180	15.7	
Starch, cellulose	$C_6H_{10}O_5$	162	15.0	
Black liquor (sulphate)			15.0	
Methanol (wood alcohol)	CH_3OH	32	15.0	
Rice straw, hulls or husk			14.0	
Peat			8.6–18.6	
Wood (50% moisture)			10.2	
Carbon monoxide	CO	28	10.11	10.11
Sulphur	S	32	8.97	8.97
Wood (80% moisture)			4.1	

Notes
(1) Major constituent of natural gas.
(2) A mixture of octane, heptane and other hydrocarbons.
(3) Aviation fuel, heating oil, oil for cooking.
(4) Sugar, starch.

Source: Avallone and Baumeister 1987

Organic waste (2)

Nearly all organic waste fuels are cellulosic in character, and the heating value is a function of carbon content. On a moisture-free basis, the heating values can be estimated at 18.8 MJ kg^{-1}; more resinous materials having about 21 MJ kg^{-1}. Table 5.11 lists some examples.

Table 5.11 Examples of organic waste

By-product fuel	Calorific value MJ/kh (dry)[1]	% moisture as received	Latent heat MJ/kg[2]	'WET' calorific value MJ/kg wet matter
Coffee grounds	23.26	65	1.63	6.51[3]
Cattle manure	17.2	50–75	1.25–1.88	7.35–2.42
Municipal refuse	22.1	43	1.15	11.45
Rice straw or hulls	13.9	7	0.18	12.75
Wheat straw	19.77	4	0.10	18.88
Corncobs	21.63	10	0.25	19.22

Notes
(1) 1 J/kg = 4.3×10^{-4} BTU/lb
2326 J/kg = 1 BTU/lb
1 BTU/lb = 0.002326 MJ/kg
(2) The latent heat of vaporisation of water is 2.5 MJ/kg.
(3) 1 kg of wet material contains 35% of dry material, 0.35 kg having a calorific value of $23.26 \times 0.35 = 8.14$MJ/kg, together with 0.65%, 0.65 kg of water which must be boiled off using $0.65 \times 2.5 = 1.63$ MJ, resulting in an overall calorific vallue of $8.14 - 1.63 = 6.51$ MJ/kg of wet material.

Combustion calculations

Combustion of methane in air

Air contains 21% O_2 and 79% N_2. Thus, for complete, or *stoichiometric*, combustion, the exhaust gas constituents are calculated as follows:

	CH_4	$+ 2(O_2$	$+ (79/21)N_2)$	$= CO_2$	$+ 2H_2O$	$+ (2 \times 79/21)N_2$
kg	16	+ 64	+ 7.52 × 28	= 44	+ 36	+ 7.52 × 28
kg	1	+ 4	+ 13.16	= 2.75	+ 2.25	+ 13.16
mols	1	+ 2	+ 7.52	= 1	+ 2	+ 7.52
m^3	1	+ 2	+ 7.52	= 1	+ 2	+ 7.52

Thus 1 kg of methane requires 17.16 kg of air for complete combustion and produces 18.16 kg of waste products: 2.75 kg of CO_2, 2.25 kg of water and 13.16 kg of hot nitrogen. From Table 5.11, 55.85 MJ of heat is released in the process, of which 5.63 MJ is in the water vapour produced. If this water condenses, the heat of condensation (5.63 MJ) is not subsequently available. Thus the combustion of 1 kg of methane provides 55.85 MJ and 2.75 kg of CO_2, corresponding to 0.049 kg CO_2 per MJ of energy released.

1 m^3 of methane requires 9.52 m^3 of air for complete combustion.

Thermal efficiency

% heat lost in the flue gas = 100% – efficiency %.

The thermal efficiency of combustion is usually based upon the net calorific value of the fuel and is defined as:

$$\text{Efficiency} = (\text{heat released} - \text{heat in flue gas})/\text{heat released}$$

Excess air

It is often necessary to supply a greater amount of air than the stoichiometric requirement so that complete combustion can be assured in practical furnaces. This has the effect of cooling the combustion products.

For stoichiometric combustion:

$$CH_4 + 2(O_2+(79/21)N_2) = CO_2 + 2H_2O + (2\times 79/21)N_2$$

If there is a % of excess air present, A = a/100. Then:

$$CH_4 + 2(1+A)(O_2+(79/21)N_2) = CO_2 + 2H_2O + 2AO_2 + 2((1+A)79/21)N_2$$

5.3 EMISSIONS TO AIR

Chronology: energy-related emissions to air

Table 5.12 lists examples of some of the major EC Action Programmes and Directives, UK Acts and Regulations relating to energy and emissions to air. The list is not intended to be exhaustive.

Prescribed substances

Release into the air (SCHEDULE 4).

- oxides of sulphur and any sulphur compounds
- oxides of nitrogen and any nitrogen compounds
- oxides of carbon
- organic compounds and partial oxidation products
- metals, metalloids and their compounds
- asbestos (suspended particulate matter and fibres), glass fibres and mineral fibres
- halogens and their compounds
- phosphorous and its compounds
- particulate matter

Table 5.12 EC Action Programmes and Directives, UK Acts and Regulations relating to energy and emissions to air

Year	Type	Development
1906	ACT	Alkali, etc. Works Regulation Act
1936	ACT	Public Health Act
1947	ACT	Town and Country Planning Act
1956	ACT	Clean Air Act
1961	ACT	Public Health Act
1968	ACT	Clean Air Act
1971	REG	Clean Air (Emission of Grit and Dust from Furnaces) Regulations
1972	ACT	Local Government Act
1973	ECAP	First EC Environmental Action Programme
1974	ACT	Control of Pollution Act
1974	ACT	Health and Safety at Work, etc. Act
1977	ECAP	Second EC Environmental Action Programme
1978	DIR	EC Toxic and Dangerous Waste Directive
1980	DIR	EC Directive on Air Quality Limits and Guide Values for Sulphur Dioxide and Suspended Particulates
1981	ECAP	Third EC Environmental Action Programme
1982	DIR	EC Directive on a Limit Value for Lead in Air
1983	REG	Health and Safety (Emissions into the Atmosphere) Regulations
1984	DIR	EC Combating of Air Pollution from Industrial Plants Directive
1985	DIR	EC Directive on Air Quality Standards for Nitrogen Dioxide
1986	ACT	The Single European Act
1987	ECAP	Fourth EC Environmental Action Programme
1988	DIR	EC Directive on the Limitation of Emissions of Certain Pollutants into the Air from Large Combustion Plant
1988	ACT	Environment and Safety Information Act
1989	ACT	Control of Pollution (Amendment) Act
1989	REG	Air Quality Standards Regulations
1989	ACT	Control of Smoke Pollution Act
1989	REG	Control of Industrial Air Pollution (Registration of Works) Regulations
1989	REG	Control of Pollution (Registers) Regulations
1990	ACT	Planning (Hazardous Substances) Act
1990	ACT	Environmental Protection Act
1991	REG	The Environmental Protection (Prescribed Processes and Substances) Regulations (replaces 1983 REG)
1991	REG	Environmental Protection (Applications Appeals and Registers) Regulations
1991	REG	Environmental Protection (Authorisation Processes) Regulations
1991	REG	EC Eco-Audit Regulation
1992	REG	EC Ecolabelling Regulation
1993	ECAP	Fifth EC Environmental Action Programme

Notes
ACT—UK Act
RES—Uk or EC Regulation
DIR—EC Directive
ECAP—EC Action Program

Prescribed substances which come under SCHEDULE 4 of the Prescribed Processes and Substances Regulations (3) relating to releases into the air are as follows:

The oxides of carbon: carbon monoxide, CO, and carbon dioxide, CO_2, emanating from fuel combustion from domestic heating, power stations, industrial processes, waste incinerators and motor vehicles. These are greenhouse gases, but also mild acid gases, forming carbonic acid.

Figure 5.2 shows the increase in carbon dioxide emissions (Mtonnes CO_2 per annum) from fossil fuel combustion over the period from 1960 to 1989 for the European Community, the United States and Canada and the rest of the world. It can be seen that the total world emission of carbon dioxide from the combustion of fossil fuels *trebled* over the 30-year period shown, corresponding to a mean annual increase of 3.7%.

The greatest percentage increase over the 30-year period considered was for those countries other than the European Community and the United States (together increasing by 78%), where annual carbon dioxide emissions increased by 318%. Total annual carbon dioxide emissions in the UK decreased from 182 Mtonnes per annum in 1970 to 159 Mtonnes per annum in 1991 (Figure 5.3). Whereas emissions from the DCP and industrial sectors decreased by 28%, this was offset by an increase of 65% from the transport sector, from 23 Mtonnes per annum in 1970 to 38 Mtonnes per annum in 1991. Figure 5.3 also indicates the restraining effects of the 1973 oil 'crisis' and the 1979 Arab–Israeli war.

Figure 5.4 shows that, in the UK, the total emission of poisonous carbon monoxide increased by 46% over the period from 1970 to 1991, this being caused by a doubling of the carbon monoxide release due to transport, from 2977 thousand tonnes per annum to 6060 thousand tonnes per annum, whilst the carbon monoxide emissions from the DCP and industrial sectors decreased by 70%, from 1468 thousand tonnes per annum to 442 thousand tonnes per annum.

The **oxides of nitrogen**, emanating from fuel combustion from domestic heating, power stations, industrial processes, waste incinerators and motor vehicles. They include the greenhouse gas nitrous oxide, N_2O, and the acid NOx gases (nitric oxide, NO, nitrogen dioxide NO_2), which form nitric acid, contributing to acid rain and precipitation. Under the action of sunlight on vehicle exhaust fumes, the NOx gases form photochemical smog as well as low-level ozone (also a greenhouse gas).

Figure 5.5 shows that, within the European Union, the greatest proportion of nitrous oxide is emitted by road transport (53%), followed by electricity generation (30.7%).

Figure 5.6 shows that total annual acid NOx emissions in the UK increased by 20%, from 2293 thousand tonnes per annum in 1970 to 2747 thousand tonnes per annum in 1991. Again, a 30% reduction in the DCP and industrial sectors was more than offset by a doubling of NOx gases emitted by transportation.

Figure 5.7 shows that the greenhouse gas nitrous oxide emanates predominantly

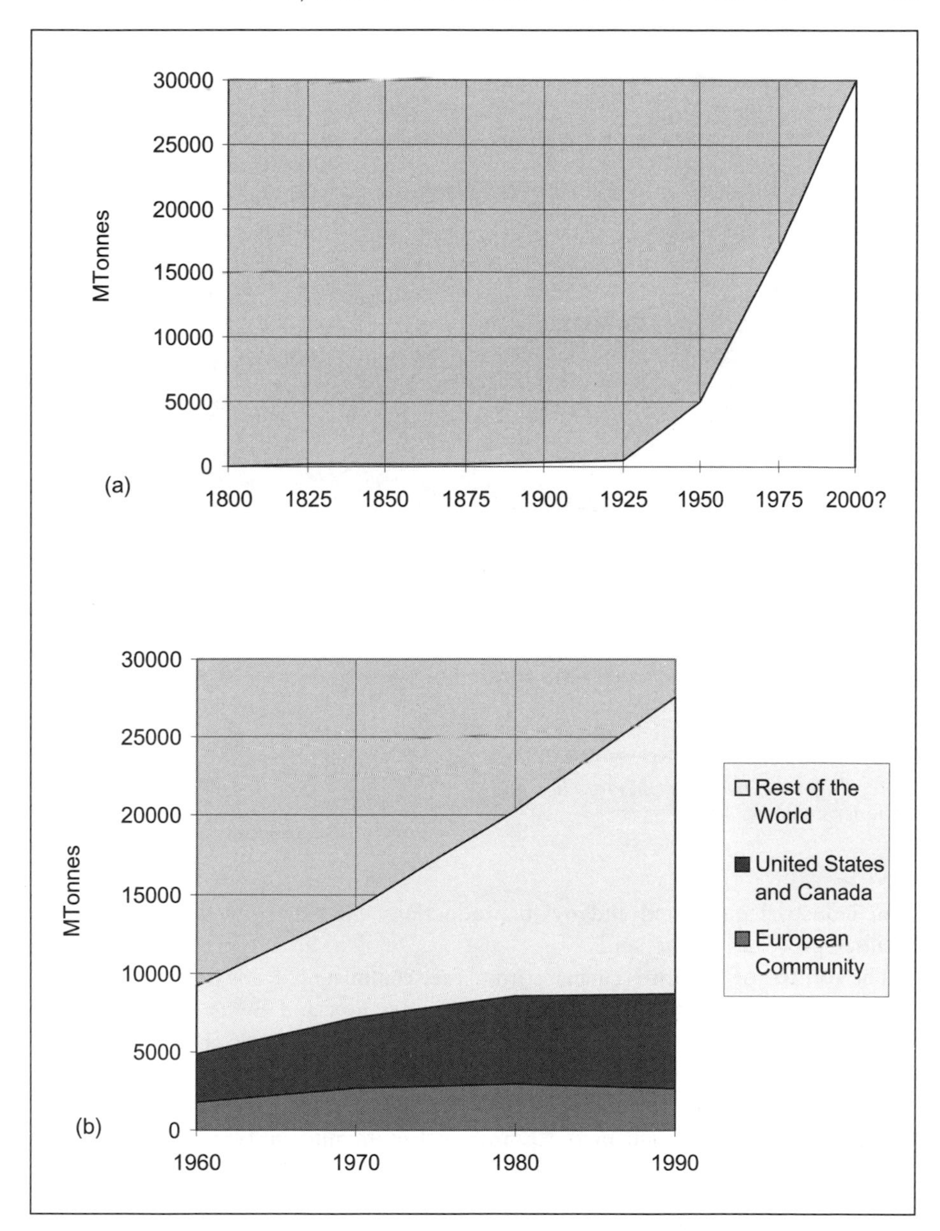

Figure 5.2 Carbon dioxide emissions from fossil fuel combustion
Source: Eurostat, *Basic Statistics of the (European) Community*, 29th Edition, 1992.

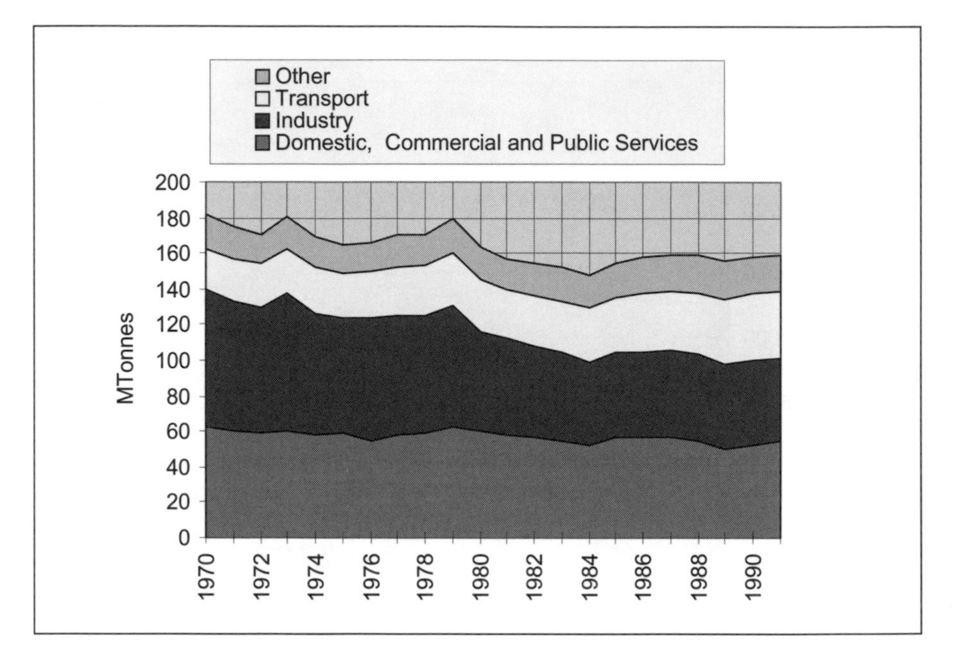

Figure 5.3 UK carbon dioxide emissions

Notes

Domestic (53–45 Mtonnes), commercial/public services (12–9 Mtonnes).

Other includes miscellaneous (19–13 MTonnes), agriculture (2–3 MTonnes), exports (1 MTonne)

Source: *Digest of Environmental Protection and Water Statistics,* UK Statistical Office, Department of the Environment, 1992.

from industrial nitric acid and nylon productions and from the denitrification of fertiliser processes in the soil.

The **oxides of sulphur** emanate from fuel combustion from domestic heating, power stations, industrial processes, waste incinerators and motor vehicles. They include the acid gas sulphur dioxide, SO_2, which forms sulphuric acid, contributing to acid rain and precipitation.

Figure 5.8 shows the breakdown of the sources of acid sulphur oxide emissions from fossil fuel combustion in the European Community in 1985. It can be seen that the greatest contributor to sulphur oxide emissions is the electricity generation sector (66.3%), followed by other industrial combustion processes.

Figure 5.9 shows that as a result of the introduction of smokeless fuels and general industrial decline, the annual rate of emissions in the UK of the acid gas sulphur dioxide decreased by 45%, from 6424 thousand tonnes per annum in 1970 to 3565 thousand tonnes per annum in 1991.

Organic compounds (volatile organic compounds, VOCs) and partial oxidation products, emanating from oil-based combustion processes, chemical

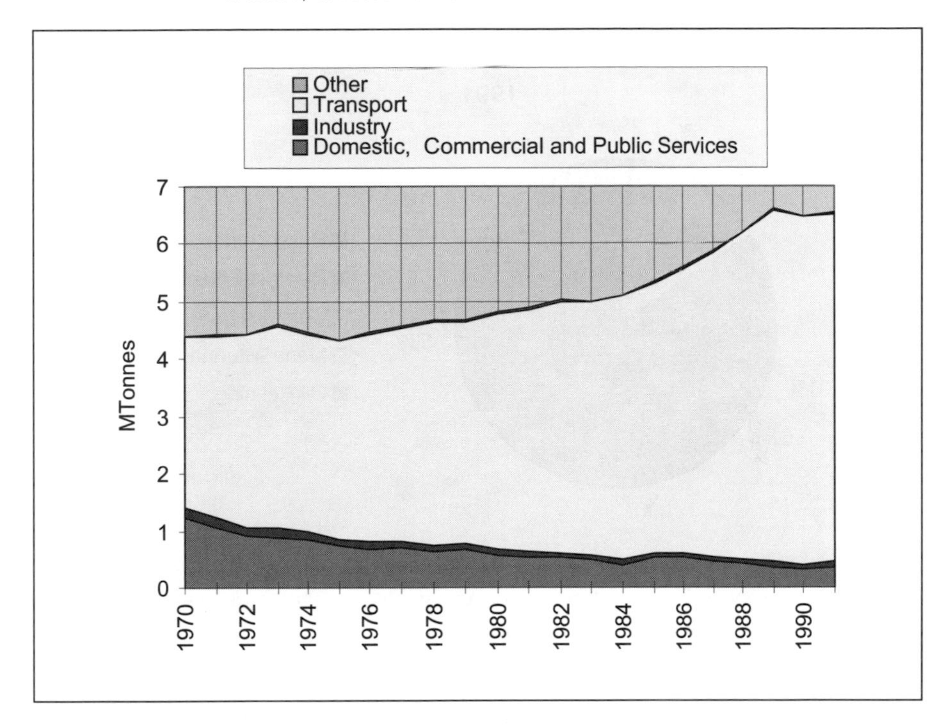

Figure 5.4 UK carbon monoxide emissions

Source: Digest of Environmental Protection and Water Statistics, UK Statistical Office, Department of the Environment, 1992.

processes, solvent and paint use and waste incinerators. They include aldehydes (organic compounds containing CHO attached to a hydrocarbon (e.g. from diesel exhausts)), polynuclear aromatic hydrocarbons (PAHs) and methane, emanating from coal mining, landfill sites, fuel combustion, natural gas leaks, ruminants, swamps, etc.

VOCs usually have an unpleasant smell, irritate the nose and eyes and can be poisonous. They consist of a large number of compounds, including hydrocarbons and oxygenated and hydrogenated organics. Some of these compounds are added to petrol, especially super unleaded petrol, to improve engine performance. Benzene (C_6H_6), from petrol combustion products and the evaporation from petrol pumps and fuel tanks, is a highly toxic carcinogen present in motor engine fuels. Figure 5.10 shows that the annual rate of emissions of VOCs in the United Kingdom increased by 22%, from 2187 thousand tonnes per annum in 1970 to 2678 thousand tonnes per annum in 1991. VOCs, besides being harmful to humans, are directly involved in the formation of tropospheric ozone, a greenhouse gas.

The quantities of exhaust gases may be calculated from the chemical constituents using the combustion equations as demonstrated in the previous section. For

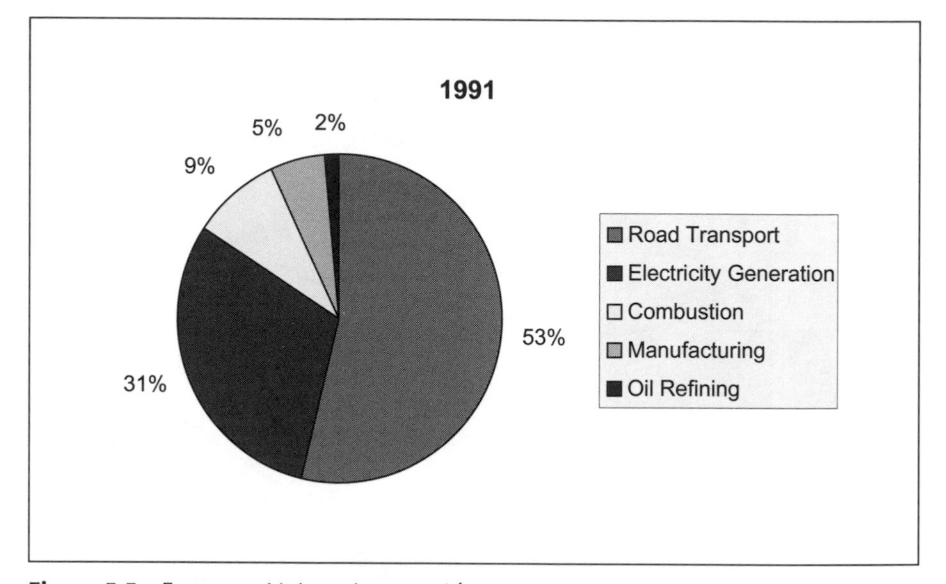

Figure 5.5 European Union nitrous oxide emissions
Source: Eurostat, *Basic Statistics of the (European) Community*, 29th Edition, 1992.

complete combustion, Table 5.13 lists the quantities of the polluting exhaust gases produced per kilogram of the various combustible substances.

Figure 5.11 shows that total UK methane emissions remained almost constant at about 3.5 Mtonnes over the period from 1970 to 1991, the most important contributors being animals, general gas leakage from the soil, emissions from landfills and coal mining. A reduction in the latter from 1.5 Mtonnes/annum in 1970 to 0.77 Mtonnes per annum in 1991 was offset by increases in methane emissions from general gas leakage, leakage from natural gas collection and emissions from landfills.

Other substances emitted to air

General matter: airborne particulate matter, dust from solid fuels, smoke, fumes (especially from liquid fuels), odours, and aerosols from liquid fuels are other energy-related emissions.

Figure 5.12 shows that, in the UK, total black smoke emissions halved over the period 1970 to 1991. As a result of the introduction of smokeless zones and smokeless fuels, the rate of black smoke emissions from the DCP and industrial sectors decreased by 75%, from 857 thousand tonnes per annum to 217 thousand tonnes per annum.

Yet black smoke emitted from the transport sector doubled, from 109 thousand tonnes per annum to 214 thousand tonnes per annum over the period considered.

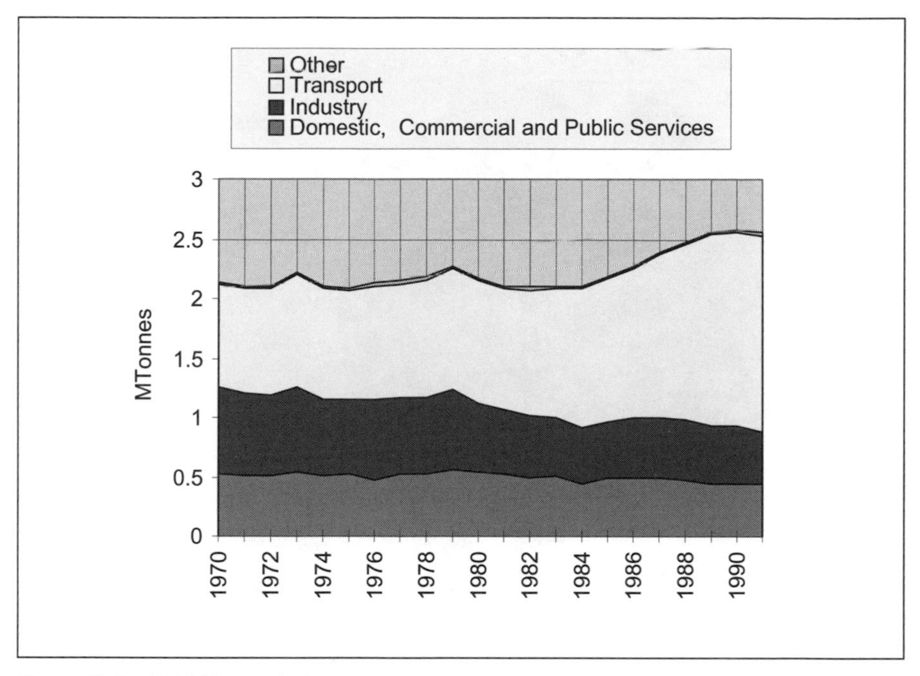

Figure 5.6 UK NOx emissions

Source: *Digest of Environmental Protection and Water Statistics*, UK Statistical Office, Department of the Environment, 1992.

The substantial increase in the number of diesel-fuelled vehicles on UK roads caused this increase. Whilst diesel engines are more efficient, producing less carbon dioxide per vehicle-kilometre and helping to combat global warming, they produce greater quantities of black smoke and other local pollutants than do petrol-fuelled vehicles.

Airborne metals, metalloids and their compounds are also emitted to air.

The rise of road transport in the UK

The UK as a whole is attempting to reduce the rates of release of polluting gaseous releases to air. The previous sections have shown that, in general, all sectors of energy utilisation have succeeded in accomplishing this to a degree, except the transport sector, which has been responsible for:

- a rise in the annual rate of energy consumption of 72% (0.74e12 MJ/annum) in 1991, compared with that used in 1970

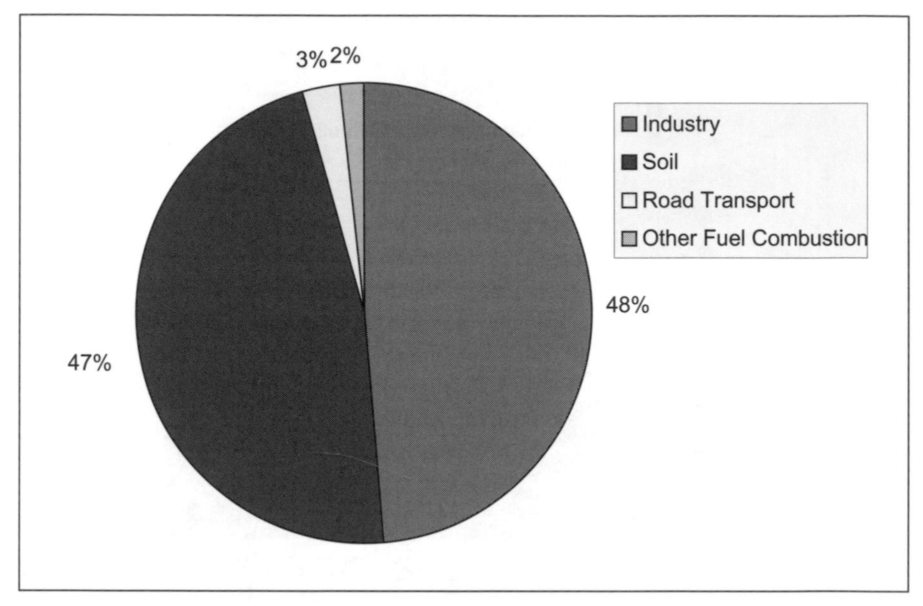

Figure 5.7 UK nitrous oxide emissions

Source: *Digest of Environmental Protection and Water Statistics*, UK Statistical Office, Department of the Environment, 1992.

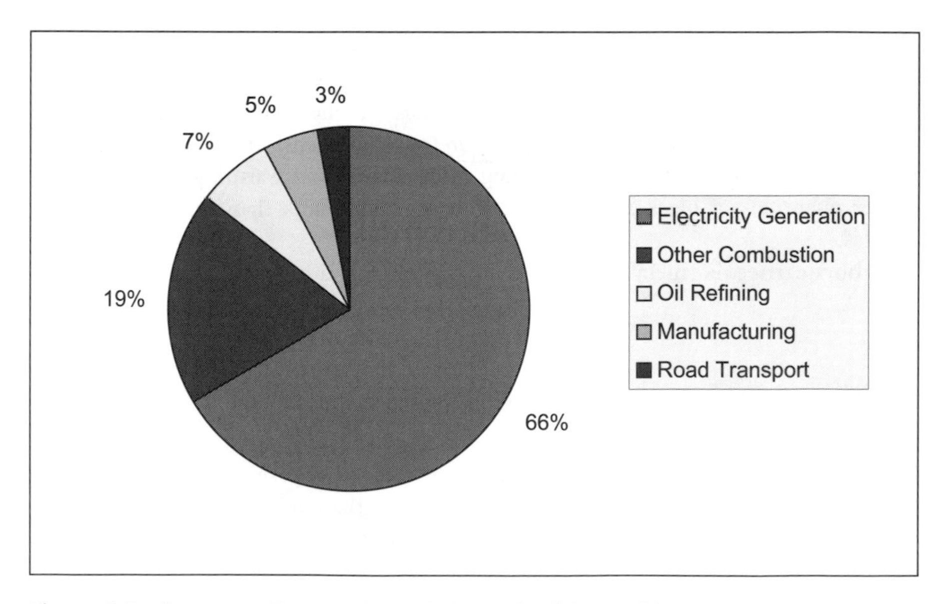

Figure 5.8 European Community emissions of sulphur oxides

Source: Eurostat, *Basic Statistics of the (European) Community*, 29th Edition, 1992.

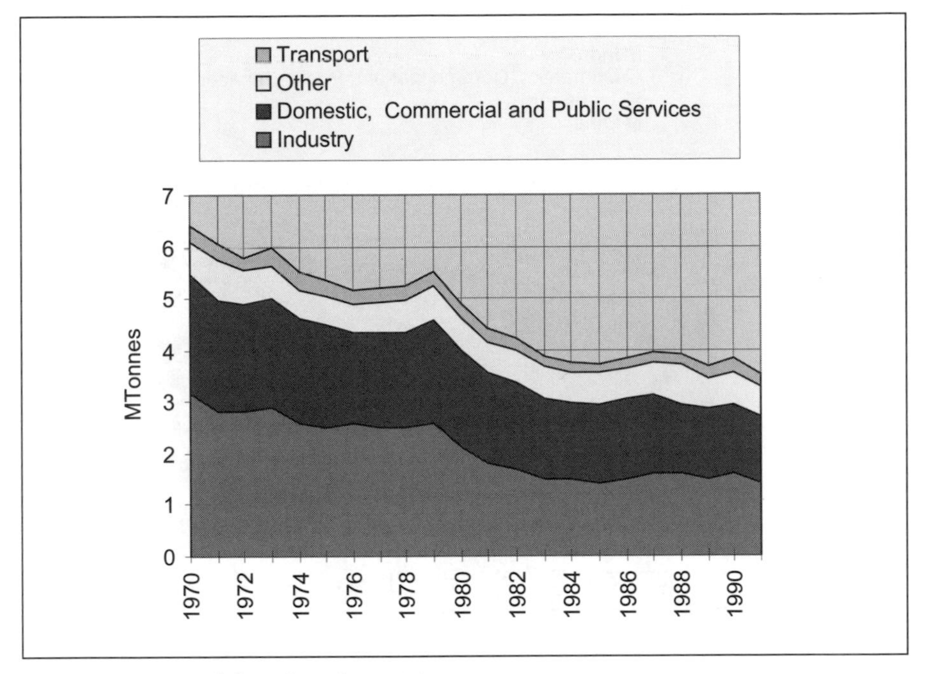

Figure 5.9 UK sulphur dioxide emissions

Source: Digest of Environmental Protection and Water Statistics, UK Statistical Office, Department of the Environment, 1992.

- a 96% increase in black smoke emitted of 105 thousand tonnes/annum in 1991, compared with that emitted in 1970
- a 103% increase in carbon monoxide emissions of 3083 thousand tonnes/annum in 1991, compared with that emitted in 1970
- a 65% increase in carbon dioxide emissions of 15 000 thousand tonnes/annum in 1991, compared with that emitted in 1970
- a 92% increase in NOx emissions of 789 thousand tonnes/annum in 1991, compared with that emitted in 1970
- a 63% increase in VOC emissions of 518 thousand tonnes/annum in 1991, compared with that emitted in 1970

The main reason for all these increases is the inexorable rise in the volume of transport. Figure 5.13a indicates the 333% growth (at 3.33% per annum) of the transport sector in the UK over the period from 1950 to 1990. Whilst passenger transportation by public transport declined by 66%, passenger transportation by cars increased over tenfold (at 6% per annum) over the period. The amount of road traffic (vehicle-kilometres) increased by 48% (at 4% per annum) over the 10 years from 1981 to 1991 (Figure 5.13b).

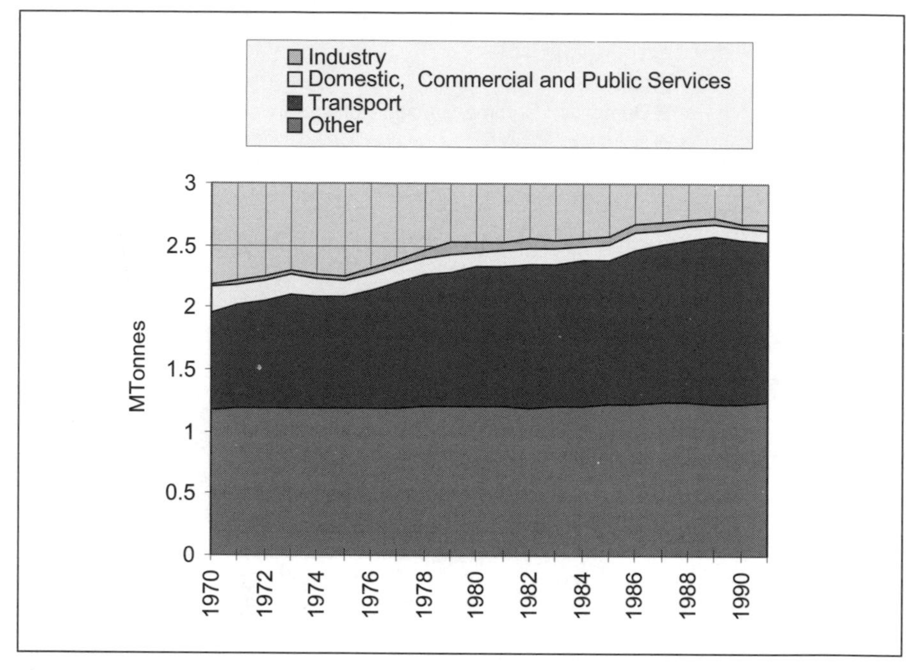

Figure 5.10 UK emissions of volatile organic compounds (VOCs)

Source: *Digest of Environmental Protection and Water Statistics*, UK Statistical Office, Department of the Environment, 1992.

As a result, the UK annual consumption of motor fuel increased by 43%, from 24.2 Mtonnes per annum in 1981 to 34.7 Mtonnes per annum in 1991 (Figure 5.13c). As a result of the introduction of lead-free petrol, lead emissions from petrol fell dramatically, from 8 thousand tonnes per annum in 1975 to 1.8 thousand tonnes per annum in 1992 (Figure 5.13d).

Normalised transport figures

The total pollution produced by U.K. transport rose by 67% from 30 Mtonnes/annum in 1970 to almost 50 Mtonnes/annum in 1991 (Figures 5.13e–j). For every tonne of motor fuel combusted, more than a tonne of pollutant is emitted. Overall pollution from UK traffic amounted to 501 kg per capita of population in 1970. By 1991, this had risen by 69% to 847.6 kg per capita. This corresponds to about 600 m^3 of pollution per person, a volume 12 000 times a typical person's own volume.

Figure 5.14 normalises these figures in terms of vehicle-kilometres and shows that the overall increase of polluting emissions from transport is due to the increase in the number of vehicles. The production of more efficient engines, the rise of diesel-fueled systems and the introduction of catalytic converters have contributed

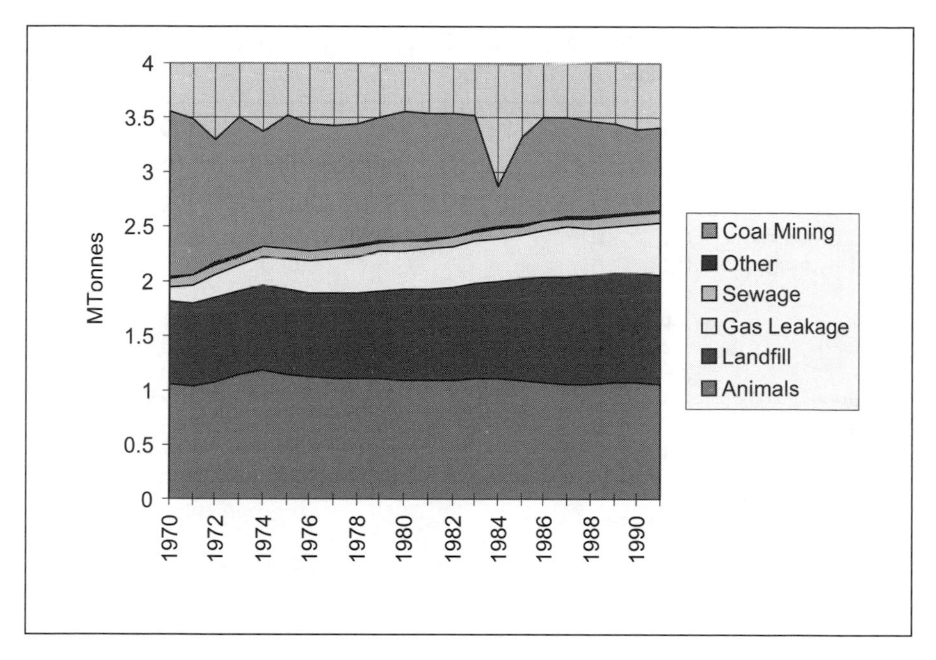

Figure 5.11 UK methane emissions

Notes

Animals:	Cattle (1001–787 Mtonnes)
	Sheep (160–261 Mtonnes)
	Other animals (17–15 Mtonnes)
Gas leakage:	General (381–409 Mtonnes)
	Offshore oil and gas (10–99 Mtonnes)
Coal mining:	Deep-mined coal (1526–764 Mtonnes)
	Open-cast coal (3–7 Mtonnes)

Source: *Digest of Environmental Protection and Water Statistics*, UK Statistical Office, Department of the Environment, 1992.

to a 10% reduction (to 90 gm per vehicle-kilometre) in the total emissions per vehicle-kilometre. The exception to this general trend is that emissions of black smoke (emanating from diesel engines) have increased by 21% from 0.43 gm per vehicle kilometre to 0.52 gm per vehicle kilometre.

Assuming that an average vehicle covers 18 000 kilometres in a year, it produces 1.8 tonnes of pollution each year, over twice its own weight and possibly 20 times its own weight of pollution in a 10-year life.

'Epidemic' rise in asthma cases

According to a House of Commons report (5), reported cases of asthma in the UK rose by a massive 160%, from 18 incidences reported per 100 000 population in

Table 5.13 Quantities of the polluting exhaust gases produced per kilogram of various combustible substances

1 kg of substance	kg CO_2	kg SO_2	kg H_2O	Gross CV MJ/kg	Net CV MJ/kg	kg CO_2/ MJ net CV	kg SO_2/ MJ net CV
Gaseous hydrogen	0	0	9	143.4	120.9	0	0
Liquid hydrogen	0	0	9	116.5	94.0	0	0
Sulphur	0	2	0	9	9	0	0.22
Carbon	3.67	0	0	32.8	32.8	0.11	0
Coke	3.22	0.0075	0.052	30.5	30.5	0.105	0.00025
Anthracite	3.05	0.023	0.055	29.9	29.9	0.102	0.00077
Bituminous coal	3.15	0.039	0.056	33.2	33.1	0.095	0.0012
Methane	2.75	0	2.25	55.8	50.2	0.055	0
Heavy fuel oil	3.14	0.08	0.27	42.5	42.3	0.074	0.0019

Notes
CV ≡ Calorific Value

Ranking kg CO_2/MJ net CV

Best	
Gaseous hydrogen	0
Liquid hydrogen	0
Sulphur	0
Methane	0.055
Heavy fuel oil	0.074
Bituminous coal	0.095
Anthracite	0.102
Coke	0.105
Carbon	0.11
Worst	

Ranking kg SO_2/MJ net CV

Best	
Gaseous hydrogen	0
Liquid hydrogen	0
Carbon	0
Methane	0
Coke	0.00025
Anthracite	0.00077
Bituminous coal	0.0012
Heavy fuel oil	0.0019
Sulphur	0.22
Worst	

1983 to 47 incidences reported per 100 000 population in 1992, an annual rate of increase of 11%. One in ten children was affected in 1992. The report blames this increase in part on fumes from vehicles and calls for extensive monitoring of air pollution, particularly in cities.

Pollution from UK industry

Figures 5.15a–c show that total pollution produced by the UK industrial sector reduced by 17% from 58 Mtonnes per annum in 1981 to 48 Mtonnes per annum in 1991. Whilst some of this reduction may have resulted from greater energy-use efficiency, coupled with exhaust gas clean-up, the bulk of this decrease is almost certainly due to industrial decline, the number of unemployed persons increasing from around 500 000 in the late 1970s to over 3 million in 1993.

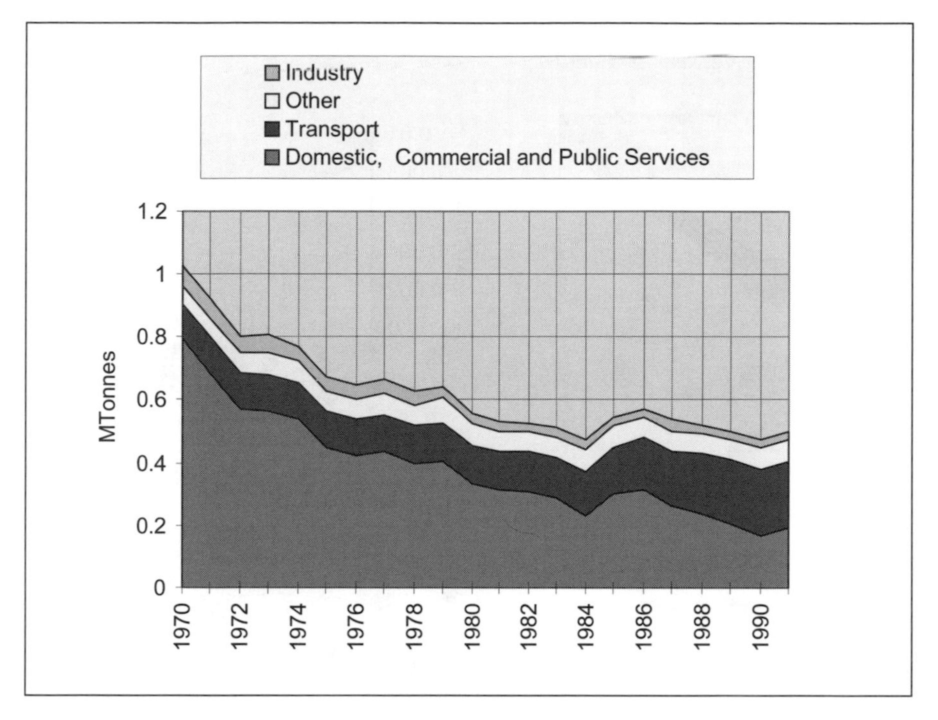

Figure 5.12 UK emissions of black smoke

Source: *Digest of Environmental Protection and Water Statistics*, UK Statistical Office, Department of the Environment, 1992.

Non-energy-related emissions to air

Prescribed substances which come under SCHEDULE 4 of the Prescribed Processes and Substances Regulations (3) relating to releases into the air are:

- Asbestos (suspended particulate matter and fibres).
- Glass fibres and mineral fibres.
- Halogens and their compounds (halogenated fluorocarbons are ethane- or methane-based compounds. Chlorofluorocarbons (CFCs—greenhouse gases and also ozone layer destructors) are a major part of this group).

Figure 5.17 shows that total CFC emissions halved over a period of just four years, from 62.8 thousand tonnes per annum in 1986 to 31.3 thousand tonnes per annum in 1989, due almost entirely to the phasing out of CFC-based aerosol propellants, replacing these with propane-based (VOC) and other systems, although the use of CFCs for foam-blowing and solvents also decreased somewhat. The total amount of CFCs employed in refrigerators and air conditioning plant increased by 41% from 6.9% in 1986 to 9.7% in 1989, reflecting a steady rise in the number of cooling systems commissioned.

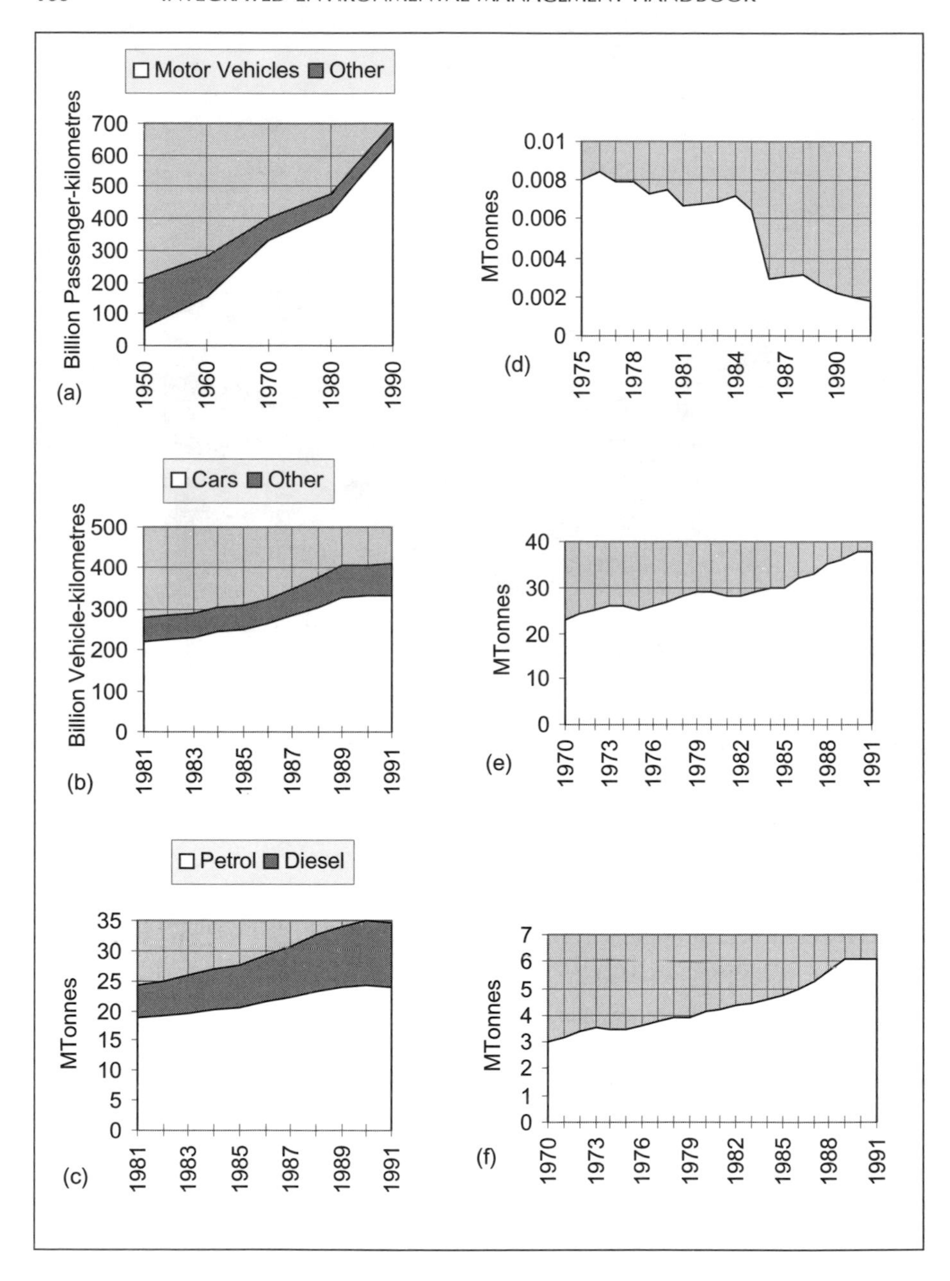
Motor Vehicles
Other
Billion Passenger-kilometres
(a)
Cars
Other
Billion Vehicle-kilometres
(b)
Petrol
Diesel
MTonnes
(c)
MTonnes
(d)
MTonnes
(e)
MTonnes
(f)

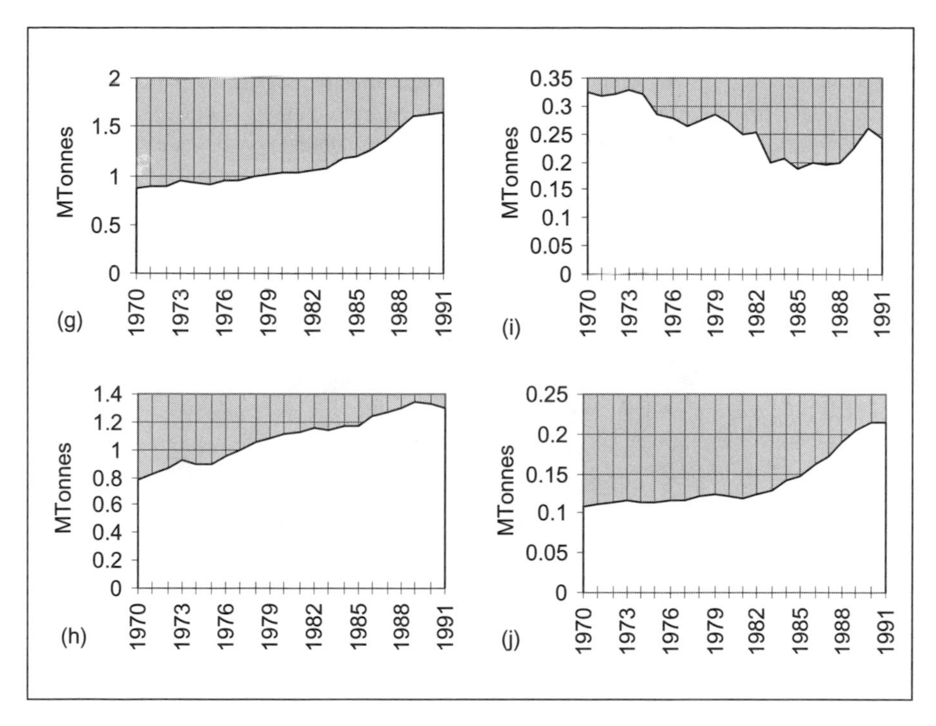

Figure 5.13 (a) UK transport; (b) UK road traffic; (c) UK motor fuel; (d) Pollution produced by transportation in the UK—lead emissions from petrol; (e) Pollution produced by transportation in the UK—carbon dioxide; (f) Pollution produced by transportation in the UK—carbon monoxide; (g) Pollution produced by transportation in the UK—NOxs; (h) Pollution produced by transportation in the UK—VOCs; (i) Pollution produced by transportation in the UK—sulphur dioxide; (j) Pollution produced by transportation in the UK—black smoke (Mtonnes)

Source: *Digest of Environmental Protection and Water Statistics*, UK Statistical Office, Department of the Environment, 1992.

- Metals, metalloids and their compounds.
- Organic compounds and partial oxidation products.
- Phosphorus and its compounds.

Other substances released into the air are:

- **General matter**: airborne particulate matter, dust, fumes, odours, aerosols (minute liquid droplets or solids of particle size up to 100 μm suspended in the air).
- **Gases and vapours**: ammonia, benzyl chloride, bromine, carbon tetrachloride (greenhouse gas), chlorine (Cl) (an irritant with severe effects on the lungs, as well as an ozone layer destructor), chloroform, chlorophene (a dangerous carcinogen), dioxine, fluorides, fluorine (F) (the most reactive element known and highly poisonous), halogenated fluorocarbons, hydrogen chloride (HCl),

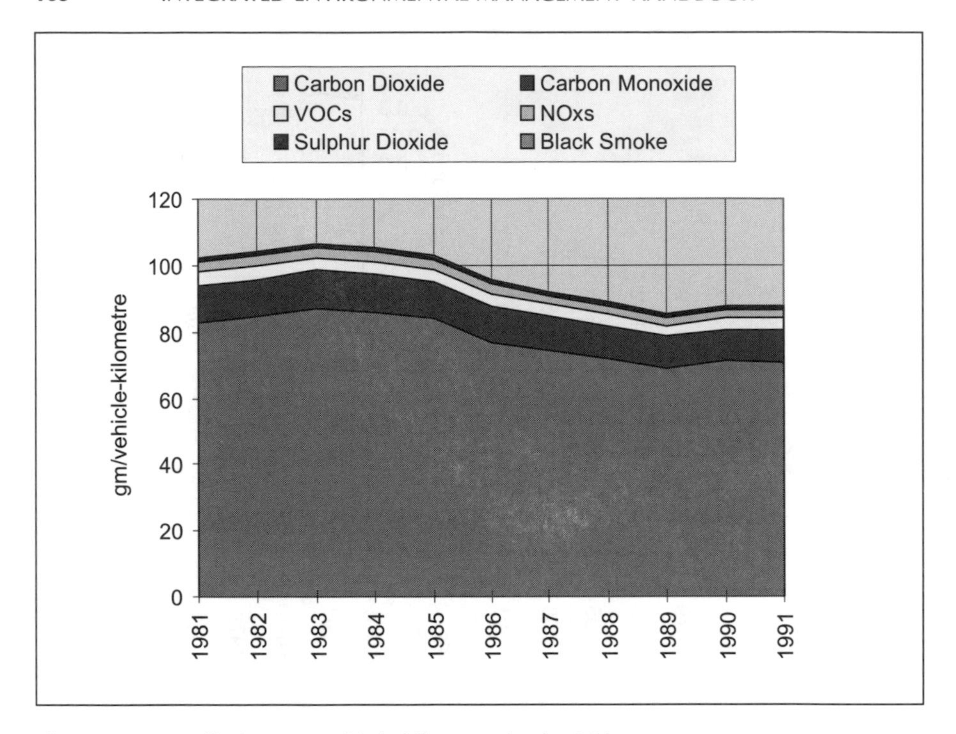

Figure 5.14 Pollution per vehicle-kilometre in the UK
Source: *Digest of Environmental Protection and Water Statistics*, UK Statistical Office, Department of the Environment, 1992.

hydrogen fluoride, the evil-smelling hydrogen sulphide and mercaptans (sulphur compounds produced at sewage works), the greenhouse gas and mild acid gas methane, methyl bromide, methylene chloride (dichloromethane), peroxyacetyl-nitrate (PAN), solvents (greenhouse gases) such as trichloroethane and trichloro-ethylene, vinyl chloride and volatile organic compounds (VOCs), including benzene.

- **Airborne metals**: aluminium, antimony, arsenic, barium, beryllium (an extremely dangerous hazardous pollutant), boron, cadmium, calcium, chromium and its compounds, cobalt, copper, cyanide, fluoride, hydrogen bromide, iron, lead, magnesium, manganese, mercury, nickel, phosphorus, plutonium, potassium, selenium, silicon, silver, sodium, tin, vanadium and zinc.

5.4 ENERGY MANAGEMENT PROCEDURE

Energy management involves the systematic quantification of energy-related activities within a system boundary, e.g. a site, a region or a product.

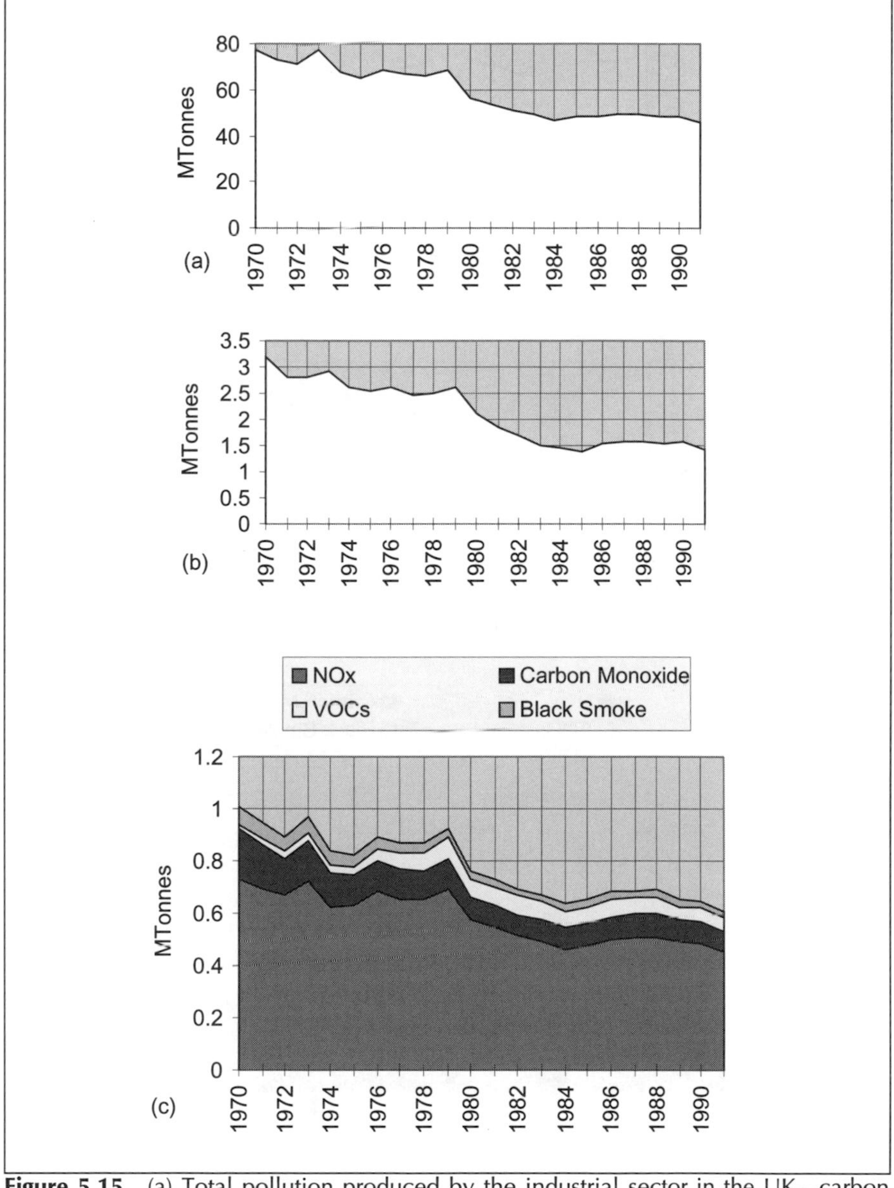

Figure 5.15 (a) Total pollution produced by the industrial sector in the UK—carbon dioxide; (b) Total pollution produced by the industrial sector in the UK—sulphur dioxide; (c) Total pollution produced by the industrial sector in the UK—other compounds

Source: *Digest of Environmental Protection and Water Statistics*, UK Statistical Office, Department of the Environment, 1992.

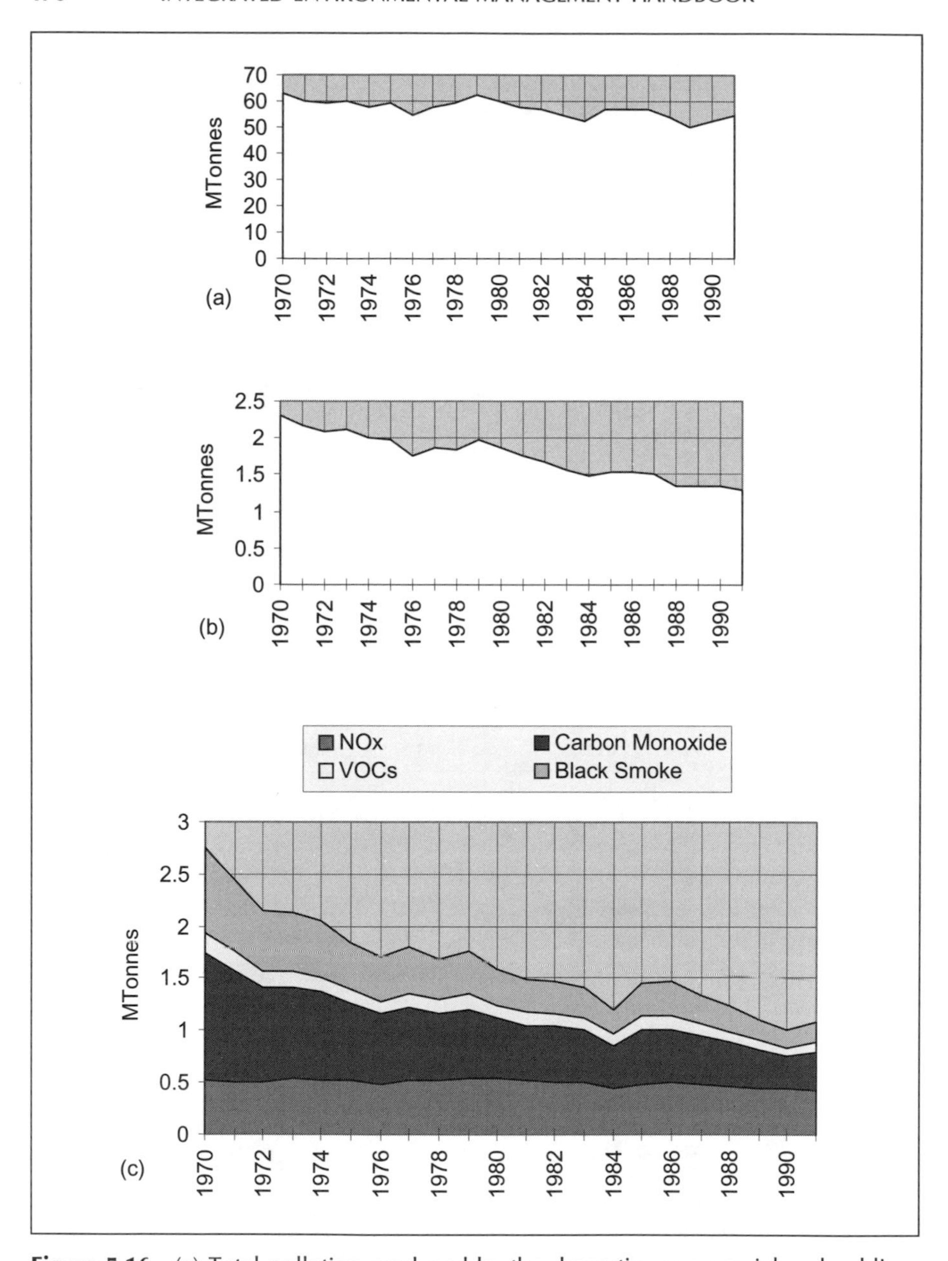

Figure 5.16 (a) Total pollution produced by the domestic, commercial and public sector in the UK—carbon dioxide; (b) Total pollution produced by the domestic, commercial and public sector in the UK—sulphur dioxide; (c) Total pollution produced by the domestic, commercial and public sector in the UK—other compounds

Source: *Digest of Environmental Protection and Water Statistics*, UK Statistical Office, Department of the Environment, 1992.

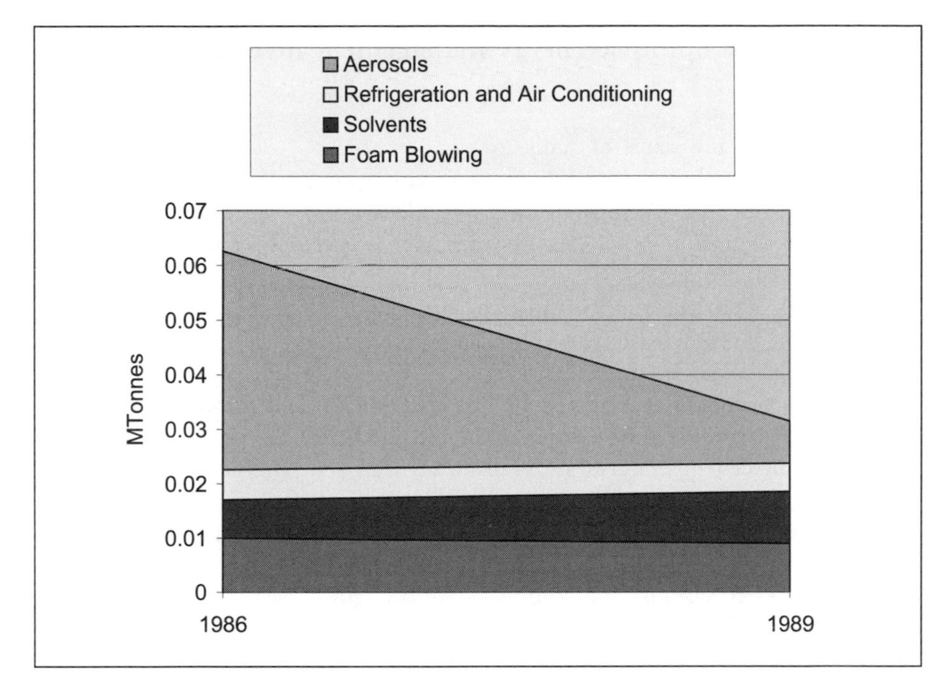

Figure 5.17 Total emissions of CFCs in the UK

Source: Digest of Environmental Protection and Water Statistics, UK Statistical Office, Department of the Environment, 1992.

An **energy account** (previously termed an energy audit but now renamed to avoid confusion with the energy aspects of environmental auditing) of a site is a balance sheet of energy *inputs*, *outputs* and *throughputs* flowing through the site boundary.

The *input* side consists of an analysis of fuel and electricity bills for a representative recent annual period.

The *output* side details the ultimate energy rejection to the external environment, mainly via heat transmission through a building's fabric and ventilating air. The data is obtained from a site energy survey.

Analyses of the *throughputs* may require microaudits, or energy balances over individual items of plant and equipment, such as furnaces, boilers, refrigeration systems, steam autoclaves, compressors, etc. to ascertain operating efficiencies and to identify where sundry gains occur.

The fundamental equation for an energy audit is as follows:

Fuel energy input
= energy losses during combustion + energy losses during conversion
+ energy losses during distribution + energy losses during utilisation
+ energy losses from utilisation

The 39 steps to optimise energy management investments at a site (6)

1. Identify site and scope of audit.
2. Obtain fuel and electricity bills for a recent representative year.
3. Obtain degree-days/temperatures for the location.
4. Identify thermal or other changes that have occurred over or since the year under examination.
5. Examine electrical supply capacity, maximum demand variations and all tariffs.
6. Feed data to a spreadsheet.
7. Convert quantities to common units, to kg of CO_2 and carbon.
8. Convert degree-days to mean monthly temperatures during operating hours unless direct temperatures are available.
9. Plot kVA versus degree-days.
10. Plot kVA versus mean monthly temperatures during operating hours.
11. Correlate with linear regression and quantify scatter.
12. Identify electricity use for heating.
13. Plot kWh of electricity versus degree-days.
14. Quantify electricity use for heating.
15. Plot kWh of electricity versus mean monthly temperature over operating hours.
16. Correlate with linear regression and quantify scatter.
17. Check electricity use for heating.
18. Obtain annual mean furnace/boiler efficiency to discount fuel supplied to obtain heating energy.
19. Plot kWh of heating energy versus degree-days.
20. Plot kWh of heating energy (including electricity use for heating) versus monthly mean air temperatures during operating hours.
21. Correlate with linear regression and quantify scatter.
22. Obtain annual mean inside air temperature.
23. Obtain base temperature.
24. Quantify sundry gains and direct rejects of energy.
25. Obtain equation relating overall U-value to number of airchanges, n.
26. External site survey—areas and volumes and capita, UA values.
27. Deduce the number of airchanges, n.
28. Construct output analysis—ventilation and fabric transmission losses.
29. Investigate other fuel-derived effluents and stack losses.
30. Internal surveys if process plant and equipment significant.
31. Construct input, output, throughput analysis.
32. Construct energy audit balance sheet.
33. Identify options for energy conservation and money saving.
34. Analyses of options—straight rate of return.

35. Produce a ranked investment portfolio.
36. Identify option conflicts.
37. Rerank the investment portfolio.
38. Establish investment plan.
39. Set up monitoring and targeting and project management systems.

Reference (6) contains two fully worked case studies which follow these sequential analytical steps. Summary data from one of these exercises follow.

Energy accounting

Energy inputs—fuels and electricity

Table 5.14 shows the fuels and electricity delivered to the site involved for an annual period, as well as the site supply capacity and monthly maximum electricity demand. These figures are converted to common units (kWh) in Table 5.15 and into costs in Tables 5.16 and 5.17.

The summary data appear in Tables 5.18 and 5.19.

Energy outputs—rejection to the external environment

The heating energy delivered is lost to the external environment via fabric transmission through the walls, roof and base (Table 5.20).

Table 5.14 Fuels and electricity delivered, Mar 91–Feb 92 (raw units)

	Heating fuel	Electricity			
Month	therms	Dayrate kWh	Nightrate kWh	Supply capacity kVA	Maximum demand kVA
Mar	27 198	162 342	43 234	1300	640
Apr	21 214	133 245	39 876	1300	550
May	10 652	154 789	43 475	1300	550
Jun	8 614	132 456	43 098	1300	560
Jul	4 563	123 453	39 876	1300	560
Aug	5 413	121 000	37 684	1300	540
Sep	15 678	124 356	37 984	1300	550
Oct	9 919	176 899	49 693	1300	560
Nov	19 804	167 564	41 453	1300	610
Dec	14 317	132 876	32 967	1300	600
Jan	29 961	152 098	37 963	1300	670
Feb	26 598	163 876	41 856	1300	690
Totals	193 931	1 744 954	489 159		

Notes
1 therm = 29.3 kWh

Table 5.15 Fuels and electricity delivered, Mar 91–Feb 92 (common units)

	Heating fuel	Electricity				
Month	**kWh**	**Dayrate kWh**	**Nightrate KWh**	**Total kWh**	**Supply capacity kVA**	**Maximum demand kVA**
Mar	796 901	162 342	43 234	1 002 477	1300	640
Apr	621 570	133 245	39 876	794 691	1300	550
May	312 103	154 789	43 475	510 367	1300	550
Jun	252 390	132 456	43 098	427 944	1300	560
Jul	133 695	123 453	39 876	297 024	1300	560
Aug	158 600	121 000	37 684	39 876	1300	540
Sep	459 365	124 356	37 984	621 705	1300	550
Oct	290 626	176 899	49 693	517 218	1300	560
Nov	580 257	167 564	41 453	789 274	1300	610
Dec	419 488	132 876	32 967	585 331	1300	600
Jan	877 857	152 098	37 963	106 791	1300	670
Feb	779 321	163 876	41 856	985 053	1300	690
Totals	5 682 178	1 744 954	489 159	7 916 291		

Table 5.16 Fuels and electricity prices (£)

Heating fuel	0.55	£/therm
Heating fuel	0.0188	£/kWh
Dayrate electricity	0.0683	£/kWh
Nightrate electricity	0.037	£/kWh
Supply capacity	0.5	£/kVA

	MD Step 1 (500 kVA) £/kVA	**MD Step 2 £/kVA**
Mar	3.83	3.5
Apr	0.9	0.5
May	0.9	0.5
Jun	0.6	0.26
Jul	0.6	0.26
Aug	0.6	0.26
Sep	0.6	0.26
Oct	0.6	0.26
Nov	3.83	3.5
Dec	11.2	8.5
Jan	11.2	8.5
Feb	11.2	8.5

Table 5.17 Fuels and electricity delivered, Mar 91–Feb 92 (£)

Month	Heating fuel	Dayrate electricity	Nightrate electricity	Total £ for kWh	Supply capacity	Maximum demand Step 1	Maximum demand Step 2	Total demand charges	Total energy bill
Mar	14 981	11 087	1 625	27 695	650	1 915	490	3 055	30 750
Apr	11 685	9 100	1 499	22 285	650	450	25	1 125	23 410
May	5 867	10 572	1 634	18 074	650	450	25	1 125	19 199
Jun	4 744	9 046	1 620	15 412	650	300	15	965	16 377
Jul	2 513	8 431	1 499	12 444	650	300	15	965	13 410
Aug	2 981	8 264	1 416	12 662	650	300	10	960	13 623
Sep	8 636	8 493	1 428	18 557	650	300	13	963	19 520
Oct	5 463	12 082	1 868	19 414	650	300	15	965	20 380
Nov	10 908	11 444	1 558	23 912	650	1 915	385	2 950	26 862
Dec	7 886	9 075	1 239	18 201	650	5 600	850	7 100	25 301
Jan	16 503	10 388	1 427	28 319	650	5 600	1 445	7 695	36 014
Feb	14 651	11 192	1 573	27 417	650	5 600	1 615	7 865	35 282
Totals	106 824	119 180	18 392	244 397	7 800	23 030	4 905	35 735	280 132

Table 5.18 Annual breakdown of energy consumption, Mar 91–Feb 92 (kWh)

Heating fuel	5 682 178
Dayrate electricity	1 744 954
Nightrate electricity	489 159
Total electricity	2 234 113
Total	7 916 291
Heat to power ratio	2.54

Table 5.19 Annual breakdown of energy and MD costs, Mar 91–Feb 92 (£)

Heating fuel	106 825
Dayrate electricity	119 180
Nightrate electricity	18 392
MD step 1	23 030
MD step 2	4 905
Supply capacity	7 800
Total electricity	173 308
Total	280 133

Table 5.20 External survey data

External survey

Dimensions:
Length m
U-value W/m²K

Roof details:
Pitched corrugated aluminium sheet with foil-backed 100 mm plasterboard

Facade	A	B	C	D	Roof	Base
Facing	S	W	N	E		
Description	Single glazing	335 mm brick	335 mm brick	335 mm brick	See above	Concrete
Length	114	70	114	70		
Height	5	5	5	5		
Area	570	350	570	350	7980	7980
U-value	5.6	1.7	1.7	1.7	1.9	.06
UA	3192	595	969	595	15162	478.8

Notes

Total UA (minus base)	20 513 WK^{-1}
Total exposed area (minus base)	9820 m^2
Overall U-value	2.09 $Wm^{-2}K^{-1}$
Total enclosed volume V	40 000 m^3
Number of ventilation airchanges	2.72 per hr
Mean inside air temperature	21°C
Mean outside air temperature	9.84°C
Working hours/annum	8760

Preliminary energy audit

Having examined the *input* and *output* sides of energy use at the site, it is possible to construct an *input/output* energy audit:

Fuel energy input
= energy content of the gas supplied
= 5 682 201 kWh/annum

Electricity used for heating is obtained by extracting the climate-dependent fraction of the electricity supplied
= 273 760 kWh/annum

Other electrical energy input (the rest of the electricity supplied)
= 1 960 353 kWh/annum

Fuel energy losses during combustion, conversion and distribution
= (1 − furnace/boiler efficiency/100) × fuel energy input
= 1 420 550 kWh/annum

Sundry gains from electricity
= 997 790 kWh/annum

Electricity directly rejected during utilisation (including that for compressed air)
= that part of the total electricity supplied that does not end up as sundry gains
= 962 563 kWh/annum

Fabric transmission heat losses
= $UA(T_i - T_o) \times$ hours/annum/1000
= 2 006 439

Ventilation heat losses
= ventilation air flow (kg/s)

=	airchanges/hr, n	
×	enclosed volume, V (m^3)	
×	density of air	(= 1.2 kg m^{-3})
/	seconds/hour,	(= 3600)
×	specific heat of air	(= 1000 J kg^{-1} K^{-1})
×	$(T_i - T_o) \times$ working hours/annum	
/	1000	(W/kW)

= nV 1.2 1000/3600 $(T_i - T_o)$ hours/annum/1000
= $(nV/3)$ $(T_i - T_o) \times$ hours/annum/1000
= 3 510 033 kWh/annum

Neglecting the small proportion of internal sundry gains from personnel (this would be significant in, for example, theatres and restaurants), and taking a furnace/boiler efficiency of 75%.

(kWh/annum)
INPUT ENERGY
5 682 201
+ 273 760
+ 1 960 353

= 0.25 × 5 682 178
+ 962 563
+ $UA(T_i - T_o) \times$ hours/annum/1000
+ $0.33nV(T_i - T_o) \times$ hours/annum/1000
OUTPUT ENERGY

INPUT ENERGY
7 916 314

= 1 420 550
+ 962 563
+ 2.09 × 9820 × (21 − 9.84) × 8760/1000
+ 0.33 × 2.72 × 40 000 × (21 − 9.84) × 8760/1000

INPUT ENERGY
7 916 314

= 1 420 550
+ 962 563
+ 2 006 439
+ 3 510 033
OUTPUT ENERGY

(kWh/annum)
INPUT ENERGY
7 916 314

= 7 899 585
OUTPUT ENERGY

Residual = 0.2% (from rounding errors)

This balanced audit has been obtained by equating the fuel and electricity inputs from an annual set of fuel bills to the energy losses occurring during combustion, conversion and distribution (assumed here to be 25%), that part of the electricity utilisation which does not result in sundry gains to the internal environment, the heat losses by transmission through the building fabric, and the heat losses associated with ventilating air.

The fraction of annual electricity utilisation which does not result in sundry gains to the internal environment, is estimated (6) by plotting the heating energy supplied each month against mean monthly external air temperatures. Although the mean inside air temperature was maintained constant at 21°C throughout the year, no heating energy would be required when the outside air temperature exceeded 19°C, the *base temperature* for the site. Thus the internal sundry gains from electrical equipment and personnel (neglected here) are sufficient to raise the internal temperature by 2°C with respect to the outside air temperature. This results in a linear relationship relating internal sundry gains to the overall U-value and the mean number of airchanges for the building.

A second linear relationship, derived from the same graph, relates total heating energy delivered per mean annual temperature difference between internal and external environments to the overall U-value and the mean number of airchanges for the building. These two linear relationships contain three unknowns, the internal sundry gains, the overall U-value and the mean number of airchanges per hour. A combination of the latter two variables was eliminated, resulting in a value for the internal sundry gains which can be compared with the overall electricity consumption.

A relationship between the overall U-value and the annual mean number of airchanges per hour was then obtained. The former was estimated from a site survey and thence the annual mean number of airchanges could be estimated.

Pollution accounting

Having constructed the *input* side of the energy accounts, it is a simple matter to calculate the annual energy-related emissions to air from, or due to, the activities at the site.

During the year considered, the site consumed 5 682 178 kWh of natural gas and 2 234 113 kWh of electricity.

Natural gas (assumed mainly methane, CH_4, releasing 50 MJ/kg) *from combustion chemistry*:

16 kg of natural gas emits 44 kg of CO_2
1 kg of natural gas emits 2.75 kg of CO_2 (0.75 kg of carbon) and releases 50 MJ (= 14 kWh)

Thus 1 kWh is provided by 0.072 kg of natural gas, whilst 0.198 kg of carbon dioxide, containing 0.054 kg of carbon, is released to the atmosphere in the process.

So, the release of 1 kWh from the combustion of methane emits 0.198 kg of CO_2.

The site then released 1125 tonnes of CO_2 (307 tonnes of carbon) accompanying the combustion of natural gas during the year considered.

Table 5.21 Typical breakdown of the constituents of natural gas

Natural gas	% C	% H_2	% S	% N_2	% O_2	% H_2O	% CO_2
	66.0	23.0	0.0	4.0	0.0	1.5	6.0

The proposed EC carbon taxes would amount to £60/tonne of carbon emitted in the year 2000. The site would then be liable to pay an additional £18 420 (an extra 17% on the bill) for the combustion of natural gas.

Natural gas is comprised mainly of methane, but also contains small percentages of ethane, propane, butane, pentane, hexane, nitrogen, oxygen and carbon dioxide (2).

As has been seen earlier, a typical breakdown of the constituents of natural gas is given in Table 5.21.

The analysis should therefore be repeated using the exact composition of the natural gas supplied to the site.

Coal (assumed mainly carbon, releasing 28 MJ/kg) *from combustion chemistry*:

12 kg of carbon emits 44 kg of CO_2
1 kg of carbon emits 3.67 kg of CO_2 (1 kg of carbon) and releases
28 MJ (= 7.8 kWh)

Thus 1 kWh is provided by 0.128 kg of carbon, whilst 0.471 kg of carbon dioxide, containing 0.128 kg of carbon, is released to the atmosphere in the process.

Electricity

1 kWh of electricity requires at least 3 kWh of coal, or 0.384 kg of carbon, to be burnt at the power station. When combusted, this releases 1.4 kg of carbon dioxide, containing 0.384 kg of carbon, to the atmosphere.

It is not clear whether the power generator would be liable to pay carbon taxes at £60/tonne of carbon emitted for this release (and then to pass this on to the consumer).

For each kWh of electricity, this tax would amount to 2.304p/kWh of electricity. During the year considered, the site consumed 2 234 113 kWh of electricity. The site would then be liable to pay an additional £51 474, an extra 30% on the electricity bill.

Table 5.1 listed typical compositions of certain coals and these are reproduced in Table 5.22. This shows that typical coals contain much less than 100% carbon, containing also hydrogen, sulphur, nitrogen, oxygen, water and ash. Steam, sulphur oxides, nitrogen oxides and residual ash are therefore also produced during coal combustion.

Table 5.22 Typical compositions of coals

Substance	% C	% H_2	% S	% N_2	% O_2	% H_2O	% Ash
Coal-4000	44.0	3.1	0.3	0.8	4.8	3.8	43.0
Coal-4500	49.0	3.8	0.5	1.0	5.4	6.0	34.4
Coal-5500	59.0	3.2	0.4	1.0	4.9	6.0	25.2

The analysis should therefore be repeated to yield the quantities of all these pollutants using the exact composition of the coal supplied to the power generator supplying electricity to the site.

Energy throughputs

Having obtained an *input/output* audit from the annual fuel bills and the external site survey, energy *throughputs* should be investigated to ascertain direct rejects of energy to the external environment other than those losses via fabric heat transmission and ventilation (e.g. process fluid effluents, flue gases from internal furnaces, etc.).

The following additional information is then required:

- plan drawings of the buildings
- elevations of the buildings
- details of the structural components of the buildings
- specifications of the boilers and associated plant
- plans of the layout of process equipment and piping
- details of air handling units and other space heating arrangements
- details and layout of the inlet and extract ventilation fan sets

(Energy *inputs* + Energy *outputs* + Energy *throughputs*) → Energy flow charts → Energy *audit*

The final energy audit relates all *inputs*, *throughputs* and *outputs* so that the effects of introducing energy-saving measures for one activity on other activities may be quantified. *Energy saved should be tracked back to that fuel (money) saved at the boilerhouse.*

Summary data from the site considered are given in Tables 5.23 and 5.24.

An analysis of compressed air usage indicated that 470 000 kWh/annum of the electricity delivered was used to supply compressed air.

Table 5.25 gives the breakdown of energy use over the period considered, and this data is used to construct the energy and monergy audit charts (Figures 5.18 and 5.19).

In addition, £36 000 was paid in electrical maximum demand and supply capacity charges.

Table 5.23 Summary data for the site considered

Number of employees	150
Working area	8000 m^2
Enclosed volume	40 000 m^3
Hrs/annum	8760
Recommended airchanges	3/hour
Calculated airchanges	2.72/hour
Overall U-value	2.09 $Wm^{-2}K^{-1}$

Table 5.24 Percentage breakdown of fabric transmission heat losses for the site considered

Component	UA (W/K)	%
Walls	2 159	10.3
Glazing	3 192	15.2
Roof	15 162	72.2
Base	479	2.3
Total	20 992	100

Table 5.25 Breakdown of energy use at the site in the representative year considered

Energy accounts	kWh	%	£	%
Ventilation losses accounted for	3 510 033	44	65 989	27
Fabric transmission losses accounted for	2 006 439	25	37 721	15
Directly rejected energy accounted for	492 563	6	30 342	12
Fuel conversion losses accounted for	1 420 550	18	26 706	11
Air compressor losses accounted for	336 990	4	20 759	9
Total losses	7 766 575	98	181 517	74
in addition				
Electrical 'added' cost of sundry gains	(997 790)	(13)	42 705	18
'Additional' cost for electrical heating	(273 760)	(3)	11 717	5
Compressed air produced	133 010	2	8 194	3
Totals	7 900 000	100	244 000	100

Energy uses and options for improvement

Having constructed the overall energy accounts, the various options for system improvement can be identified and evaluated economically. These may be classified into *demand side options* and *reject side options.*

The *demand side options* include all the ways in which energy may be saved by employing energy-conservation techniques and energy management.

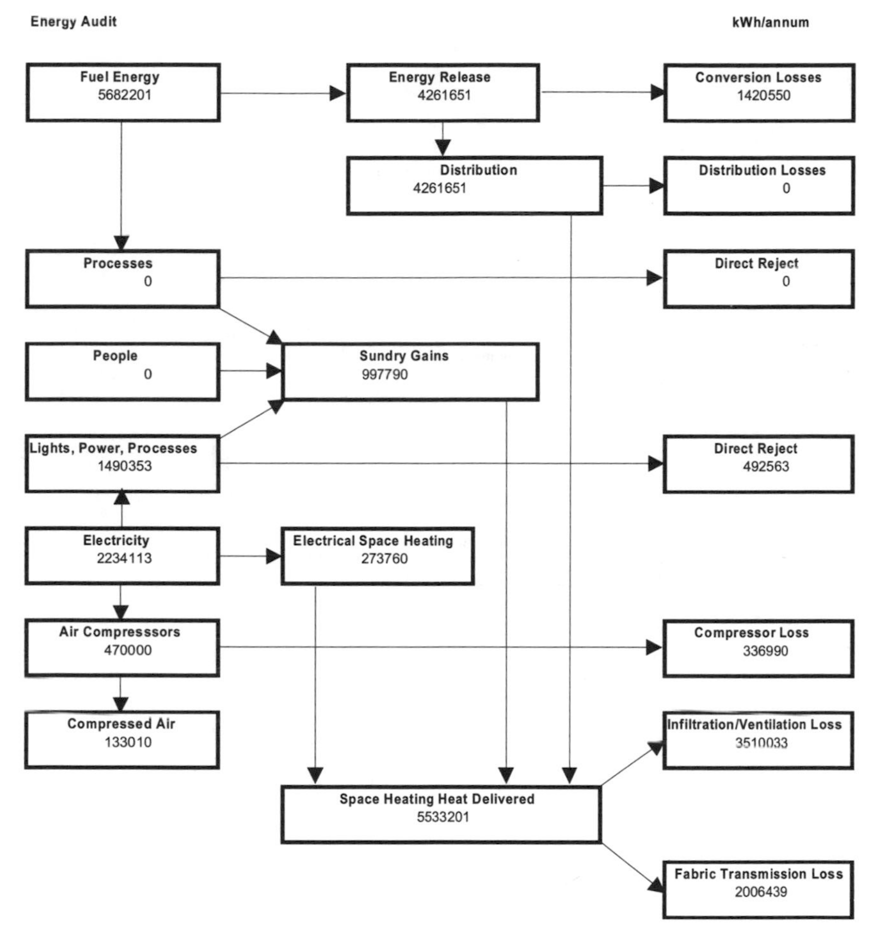

Figure 5.18 Energy audit chart

Rules for the efficient conservation of energy

General

The purposes for which expenditures of energy or materials are required should be critically examined.

As much useful work, heat or other purpose fulfilment should be extracted from a degrading energy or materials chain as is compatible with economic and other considerations. The quality, not the quantity, of energy (materials) is the subject of conservation.

Each energy operation should be examined critically and systematically in isolation and in relation to all other events occurring within the system boundary.

The manner and extent of all energy and materials use should be challenged,

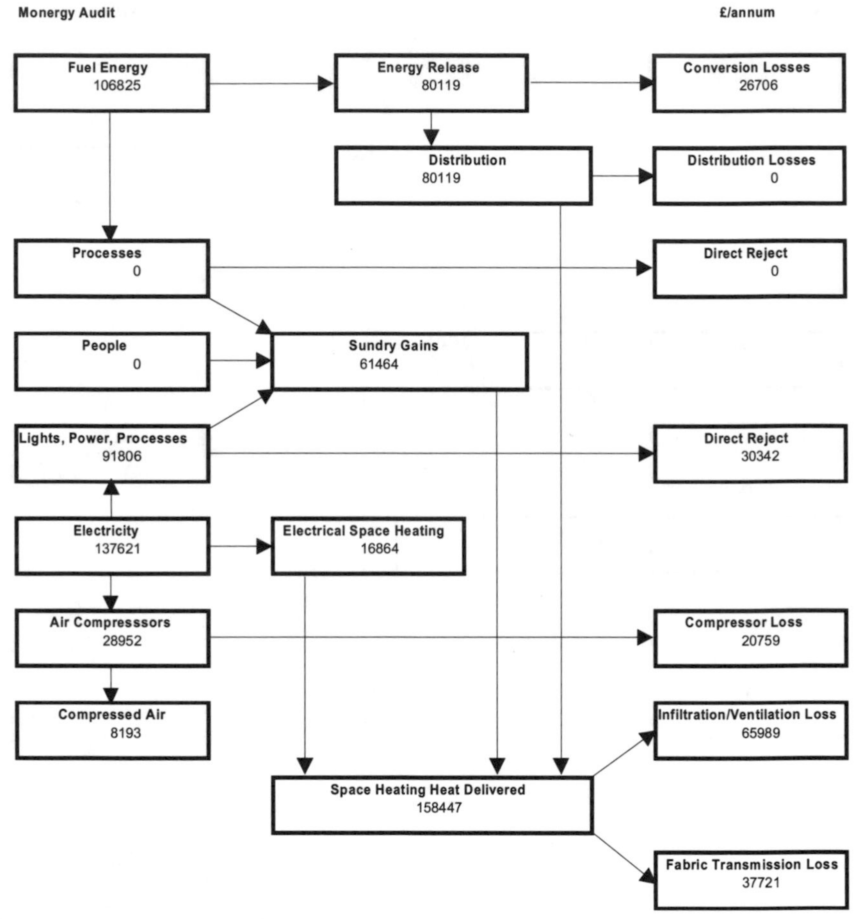

Figure 5.19 Monergy audit chart

including the appropriateness of the process method and the size of the plant involved.

Sources

Fuels and materials should be used only when and where required. Space or time delays inevitably incur losses.

Stocks should be maintained at minimum levels plus emergency reserves. Attention should be paid to the delivery, storage and handling systems. The financial, energy and materials costs of these should be assessed.

Comprehensive and accurate monitoring and metering of all energy and materials inputs, throughputs and outputs should be accomplished.

A continuous fuels log should be maintained. Procedures should be standardised. Quality should be checked.

Information should be easily accessible, comprehensible and disaggregable. Attempts should be made to account for all inputs in terms of outputs. Storage areas should be made secure against loss or theft.

Plant, equipment, systems, products

All hardware should be matched to the purpose for which it is required.

All systems should be operated at rates corresponding to maximum efficiency (normally fully loaded in continuous operation). Intermittent operations and fluctuations should be avoided. Efficiency checks should be carried out frequently using standardised procedures. Plant should be selected on sensible extreme conditions.

Energy or materials cannot readily be conserved unless accurate and comprehensive measurements in consistent units are first obtained for all activities within the system boundary. Greater overall efficiency can always be achieved at the cost of additional complexity.

Greater energy efficiency always requires an expenditure of materials and vice versa (e.g. the greater the area of a heat exchanger, the more effective the transfer of heat). The 'law of diminishing returns' applies.

Side benefits and diseconomies—incidental benefits or penalties arising from each consuming activity—should be identified and carefully evaluated. Only the most efficient component branches in energy or materials utilisation chains should be adopted. Overall efficiencies are always lower than that of the most inefficient link in the chain.

Product designs should maximise lifespan, promote easy maintenance and repair, require little additional energy or materials inputs during active life, and should facilitate reuse, recycling, easy disposal and natural degradation and recycling.

Improvements

Systems should be modelled and evaluated accurately so that the cost-effectivenesses of conservation options can be compared realistically. Careful assessments of real savings should be made, including maintenance costs.

Full audits in common units should be carried out before and after improvements. Evaluations should be obtained with respect to quantities of energy and materials, financial costs of energy and materials, energy costs of energy and materials, exergy degradation and overall exergetic efficiencies. Representative periods should be adopted for these analyses.

A continuous *monitoring and targeting* procedure should be implemented.

The selection of new plant, processes, or energy (or material) conserving measures should be made not on least capital cost criterion alone, but upon the basis of least total cost over the lifetime of the system.

Random factors should be eliminated—the system should be isolated from its external environment. All leaks of fuels, materials and energy should be prevented. Attempts should always be made to reduce demand before increasing energy or materials supplies.

It should be ensured that modifications have no hidden diseconomic effects and that they comply with safety, fire and statutory regulations and codes of practice.

Energy and exergy

Energy (materials) grade and availability should be matched to the purpose for which it is required in terms of temperature, pressure, heat flux and the qualities of materials.

The choice of an energy (materials) form to suit a particular application should not be made arbitrarily. Forms of different qualities are suitable for specific applications.

High grade energy should be used only for high grade purposes, such as producing work, fuels, materials or electrical potential.

Whenever energy quality (exergy) must be degraded, attempts should be made to:

- obtain useful heat (i.e. from a temperature reduction)
- obtain useful work (i.e. from a pressure reduction)

A degrading energy chain should be made to do as much work and other useful activity as possible during the process.

The input and output grades of energy supplied to and rejected from any particular component activity should be such that the exergetic efficiency of the activity is at maximum. Energy flows should be redirected within the overall system such that the overall exergetic efficiency of the system is at maximum.

If additional energy has to be added to a degrading energy chain, it is more efficient to introduce low grade energy, preferably 'waste' heat rejected from a higher grade activity, at the lower end of the chain. For example, fuel should not be burnt at 1000°C in order to provide space heating for a room whose temperature needs to be raised only a few degrees above that of the outside air.

If energy is to be removed from a degrading energy chain, it is more efficient to withdraw this energy before it is degraded, using the external environment as a sink, at the higher end of the energy chain.

Reclamation

High grade energy (which may be 'hot' or 'cold' with respect to the environmental datum) should not be allowed to be dissipated directly to the environment. The energy rejected from a high grade process should be collected and redirected via heat exchangers (or simply fans or pumps) to be employed at another place, collected and stored to be employed at another time, or concentrated for another higher grade purpose using a heat pump or other thermal transformer, as long as these operations are economically justifiable.

The effects on the desired purpose of by-passing, deleting or moving back up degrading energy chains should be examined. Attempts should be made to introduce feedback from energy loss centres to higher grade stations in the energy flow sequence (e.g. by recycling materials, heat pumping or incinerating waste).

Attempts to reduce or reuse waste should be made before any attempts at recycling or recovery. Waste energy and materials should be reused wherever economically possible, ensuring that practical grade, time and space-matched uses have been found for the reclaimed amounts. The value of the savings must clearly exceed the cost of recovery.

Waste heat, materials and ambient energy

The use of external waste heat, materials or ambient energy (i.e. solar, wind, waves, tides, external air temperatures) should be considered before increasing the rate of usage of fuels. Solar energy is the most pollution-free power source.

Energy storage

The use of energy accumulation should be considered to:

- balance load factors
- peak lop and use off-peak electricity
- increase overall energy efficiencies of boilers and distribution systems
- harness ambient energy

Management

Great care should be taken to ensure the cleanliness, correct operation and planned systematic maintenance of all storage, release, distribution, utilisation, insulation, heat recovery, instrumentation and control systems.

Pollution

Waste and pollution should be closely monitored and minimised. Waste in all forms not only squanders human effort, energy, time and materials, but also damages the external environment and disrupts ecological harmony.

The reduction of waste is especially desirable where materials have intense availability contents or where their historical availability costs are high. Metals, glass, plastics, paper and refractories are examples of such materials.

Design improvements, which prolong the lifespan or promote the reuse or easy recycling of these energy-intensive materials, are highly desirable. Methods of waste collection, sorting and reclamation should be developed. Improved recycling techniques are needed.

Education

All personnel should be made fully aware of the energy and materials implications of their activities and decisions.

All associated personnel should be given information, demonstrations of achievements and reports of failures of 'improvement' activities.

The following list of questions has been constructed from the perspective of the situation where an independent energy auditor is examining a client's site.

Questions for the client interview

Energy management procedures

- Who is responsible for energy management?
- Position in the organisation? Reporting to?
- Full or part time?
- Qualifications and experience?
- What is done?
- Has an energy flow diagram been prepared?
- What has been achieved?

Financial practices

- Who controls the capital spending budget?
- Who controls the recurrent spending budget?
- Upon what financial criteria should cost-effectiveness calculations be performed?
- What is the period available to complete this exercise?
- Is there a list of energy-saving investments under review, ranked in order of priority, with detailed costing and cost–benefit calculations?
- If not, why not?

Comments on energy consumption

- Is energy consumption about right, too high, too low?
- What are the areas of high energy consumption?
- What tariffs are used?
- Why these?
- When were these last reviewed?
- Can off-peak tariffs be used?
- Can maximum demand be cut?
- Can power factors be improved cost-effectively?

Monitoring and recording practices

- How is energy consumption reviewed? By whom?
- When was the last review?
- How is energy consumption analysed?
- Does the analysis normalise the data with level of activity? By building, by product, by month, by year, by cost, by use activity, by sector or section?
- What units of measurement are used?
- What are the metering control arrangements?
- Is there an energy consumption forecast/budget?
- Have standards been set? (i.e. for a given task or product or building)
- Is consumption compared with previous periods, other locations, other companies or other industries?
- What are the monitoring and targeting procedures?
- Should an energy management system be installed?
- What should be the nature of this system (computational or manual)?

Personnel energy awareness

- Are details of energy consumption made known to employees?
- Are employees made aware of the need for energy conservation?
- What steps have been made to promote energy awareness via education and training, posters, etc.?

Current-energy conservation measures

- What steps have been taken to reduce energy consumption?
- What steps have been taken to cascade or recycle energy? e.g. incineration or sale of combustible of recyclable scrap or refuse, waste heat recovery from air, water or hot products?

Comments on energy inefficiencies

- Are there obvious incidents of energy wastage?

Conditions of buildings, plant and equipment

- Are insulation and draughtproofing adequate and in good repair? Roof, walls, floors, interzones, doors, windows?

Furnaces

- Is plant operating efficiently?
- Are furnaces efficiency tested?

- What are the maintenance procedures?
- What are the control arrangements?

Boilers

- Is plant operating efficiently?
- Are boilers efficiency tested?
- Is optimum blowdown maintained?
- What are the maintenance procedures?
- What are the control arrangements?

Boilerhouse auxiliaries

- Is plant operating efficiently?
- How is it examined?
- What are the maintenance procedures?

Heat distribution systems

- Are there leaks of hot water or steam?
- Are pipes and ducts adequately lagged?
- What are the maintenance procedures?

Energy storage systems

- How are storage tanks heated?
- What are their temperatures?
- Why these temperatures?
- Are storage tanks adequately insulated?
- What are the maintenance procedures?
- What are the control arrangements?

Process plant

- Is plant operating efficiently?
- Are process temperatures and pressures at the lowest essential levels?
- What are the maintenance procedures?
- What are the control arrangements?

Space heating services

- For what periods are the buildings heated?
- How is heating controlled?

- Could the temperature be reduced?
- Does the temperature vary in different zones?
- Are parts of the building being heated unnecessarily?
- What are the maintenance procedures?

Lighting systems

- Are parts of the building being lit unnecessarily?
- How is lighting controlled?
- What are the maintenance procedures?

Power and electrical services

- Is plant operating efficiently?
- What are the maintenance procedures?

Mechanical ventilation

- What is the minimum ventilation rate?
- Why this value?
- How is the building ventilated?
- How does air get in?
- How does air get out?
- Do ventilation rates vary in different zones?
- Are parts of the building being ventilated unnecessarily?
- Is there any evidence of vertical stratification?
- What are the maintenance procedures for fan sets?
- What are the control arrangements?

Air conditioning systems

- Is plant operating efficiently?
- What are the maintenance procedures?
- What are the control arrangements?

Domestic hot water systems

- Are there leaks of hot water?
- What are the maintenance procedures?
- What are the control arrangements?

Compressed air services

- What is the compressed air requirement?
- Where does inlet air to the compressor come from?
- Where is the inlet air duct sited?
- What is compressed air used for?
- What are the delivery temperature and pressure?
- What are the pressures at the points of use?
- Is there any evidence of unauthorised use or leakage?
- Is plant operating efficiently?
- What are the maintenance procedures?
- What are the control arrangements?

Refrigeration plant

- Is plant operating efficiently?
- What are the maintenance procedures?
- What are the control arrangements?

Chilled water distribution systems

- Is there evidence of leakage?
- Is insulation adequate?
- What are the maintenance procedures?

Steam plant

- Is plant operating efficiently?
- Is there obvious leakage of steam?
- Is condensate recovered?
- Are steam traps cleaned and unclogged?
- What are the maintenance procedures?
- What are the control arrangements?

Other services

Other plant

Special equipment and processes

Energy throughputs: major items

- What are the major energy-consuming items of plant and equipment?

Information should be sought under the headings listed in Table 5.26 for the major energy-consuming items of plant and equipment.

Table 5.26 Major energy-consuming items of plant and equipment

Item	Description	Power rating	Efficiency	Operating hours and loading patterns	Control arrangements	Maintenance procedures
Furnaces						
Boilers						
Boilerhouse auxiliaries						
Heat distribution systems						
Energy storage systems						
Process plant						
Space heating services						
Lighting systems						
Power and electrical services						
Mechanical ventilation						
Air conditioning systems						
Domestic hot water systems						
Compressed air services						
Refrigeration plant						
Chilled water distribution systems						
Steam plant						
Kitchens and catering						
Other services						
Other plant						
Special equipment and processes						

Checklists for energy managers

Fuels

All leaks of fuels, materials and energy should be prevented.

Coal

- Avoid manual handling
- Use recent deliveries first

Oil

- Minimise the use of trace lines
- Control the temperatures of tanks and pipelines
- Optimise atomising temperatures

Gas

- Check continuously for leaks
- Optimise storage arrangements

Electricity

- Examine tariff structure
- Meter electricity use to all sectors
- Select optimal tariffs
- Attempt to balance load factors
- Identify equipment contributing to peak demand
- Check and correct power factors
- Peak lop
- Stagger start-up times
- Reschedule peak activities
- Convert to thermal energy (heat or cold) at off-peak periods and introduce thermal storage
- Consider the use of standby generators to peak lop
- Try to use nightrate electricity, i.e. for charging the batteries of electric vehicles
- Introduce compressed air storage to:
  - peak lop
  - use off-peak electricity
- Select electrical motors so that they run at near full load
- Consider the uses for variable-speed drives
- Maximise power factors—introduce capacitances
- Switch off plant and lighting when not required
- Pay attention to lighting

- Consider the use of standby generators for peak lopping
- Consider the introduction of a *total energy* system
- Invest in an *energy management control system*

Electricity should only be used as a last resort. This statement arises from the fact that, for every kWh of electricity consumed, 3 to 4 kWh of fossil fuel is consumed at the power station. For certain industrial heating applications, however, it may be more efficient to heat by induction or microwave heating in which the heat can be precisely directed to the component to be heated.

Energy-release furnaces

For example, batch ovens, rotary kilns, tunnel kilns.

- Check and review maintenance and operating procedures
- Check conditions of plant and equipment
- Check that plant is operating efficiently
- Check the control arrangements
- Check the adequacy and operation of monitoring instruments and controls
- Check air–fuel ratios
- Carry out combustion performance tests at regular intervals
- Evaluate performance by comparing the fuel input to perform a specific task under standard conditions
- Construct an energy balance over the furnace
- Perform efficiency checks—indirect (flue gas losses)
- Check furnace insulation
- Check state of furnace lining
- Optimise insulation levels
- Look for air leaks into furnace—confounding air–fuel ratios
- Eliminate infiltration—a slightly positive internal pressure will assist this
- Use doors for as short a time as possible—consider the use of chain barriers, air curtains and the like
- Check burners and combustion conditions
- Avoid fluctuations in fuel or air supplies
- Select proper firing and control equipment
- Minimise excess air—do not produce CO unless a reducing (non-oxidising) atmosphere is required
- Optimise flame temperatures
- Avoid flame impingement (sting) on refractory surfaces
- Aim for a uniform flame distribution filling the combustion chamber

Optimise conditions to give maximum heat transfer rates by convection or radiation (or by both mechanisms according to the purpose of the furnace and the method used to achieve this purpose—drying versus sensible heating).

To maximise radiative heat transfer:

- gas and surface temperatures should be high
- the flame should be highly luminous
- the distance between flame and stock should be short
- the flame coverage of the hearth should be good

To maximise convective heat transfer:

- gas temperatures should be high
- gas velocities should be high
- stagnant pockets should be avoided by good circulation and mixing
- all heat transfer surfaces should be maintained clean

Warm waste gases rather than cold inlet air should be used for dilution (i.e. for flue gas temperature control). Maintain a full load if possible. Investigate load patterns and operating cycles.

Attempt to balance load factors and hence avoid periodic modulation or intermittent operation (switching) when much of the energy supplied heats up the furnace to the working temperature, the furnace room and the surroundings.

Use lightweight carriers—especially in highly intermittent plant or tunnel kilns. Look for heat recovery opportunities. Check possibilities for heat recovery.

Energy conversion—boilers, autoclaves and liquid heaters

- Check flow and return temperatures
- Check conditions of plant, equipment and flues
- Check steam pressures and temperatures
- Check for leaks
- Check levels of insulation
- Check insulation of hot wells
- Investigate loading schedules
- Check load patterns and operating cycles
- Investigate sequencing of modular boilers
- Check the control arrangements
- Investigate adequacies and operations of controls for:
  - start-up
  - modulation
  - sequencing of modular boilers
- Check maintenance procedures
- Carry out boiler efficiency checks, i.e. direct (e.g. steam generated/fuel supplied)
- Carry out combustion performance tests at regular intervals
- Construct an energy balance over the device

- Before increasing device capacity, seek every opportunity to reduce demand, smooth the load and increase thermal efficiency
- When demand fluctuates, use energy accumulation to smooth firing rates
- Always operate in accordance with design specifications
- Check cleanliness of heat transfer surfaces
- Ensure that correct feedwater treatment is carried out—this reduces inefficiencies due to scaling and blowdown losses
- Check blowdown arrangements
- Recover heat during blowdown if possible
- Recover as much condensate as possible
- Ascertain effects of energy-conserving measures on boiler loads and performances
- Look for heat-recovery opportunities

Boilerhouse auxiliaries

- Check pump glands for leakages
- Check motors, bearings and belts
- Check boiler feedwater treatment
- Look for heat recovery opportunities:
  - from blow-down
  - to preheat air or feedwater
  - from economisers
- Check make-up water quantities
- Check insulation of heated pipework and storage vessels
- Check all lines for leaks
- Check ventilation arrangements

Indirectly heated vessels

- Check insulation and covers
- Check heating supply conditions
- Check load patterns and operating cycles
- Check heat transfer surfaces
- Check controls
- Check steam trap operations
- Look for heat-recovery prospects
- Consider thermal accumulation for ballasting mismatches between supply and demand

Driers

- Check water quantities (too much or too little)
- Check insulation of drying plant

- Check drier gas circulation patterns
- Check load patterns and operating cycles
- Check controls and monitoring procedures
- Check waste heat-recovery operations and possibilities

Heat distribution systems

- Check conditions of plant and equipment
- Check for leaks of hot water or steam
- Leaks should be prevented
- Check amount and condition of thermal insulation on equipment and pipework
- Estimate distribution losses
- Apply optimal levels of insulation and ensure that the insulant does not become dirty, compressed, waterlogged or degraded
- Ensure that direct losses from uninsulated pipelines, heaters or hot surfaces through building boundary walls do not occur
- Insulate all hot (cold) storage tanks to optimal economic levels

Heating systems

- Check heating control arrangements
- Check building/system response to controls
- Check that temperatures and ventilation rates are not excessive
- Check supply and return temperatures
- Check conditions of plant and equipment
- Check insulation levels and conditions throughout
- Check maintenance procedures
- Specify the exact purpose for which heating is required (sensible heating versus drying)
- Analyse load profiles
- Do not overcapitalise plant—reduce the demand before increasing the amount of heat supplied
- Consider thermal accumulation to avoid maximum demand and low or intermittent fire inefficiencies
- Keep heat transfer surfaces clean
- Look for heat-recovery opportunities

Space heating

- Stick to minimum fresh air requirements
- Consider the use of obscuration meters
- Question the minimum airchange rate adopted

- Ensure that the required airchange rate is attained in working areas, but do not airchange volumes where fresh air changes are not required (e.g. roof spaces, storage areas, unoccupied rooms or regions)
- When minimum airchange rates are being achieved, consider the introduction of waste heat recovery
- Isolate and vent-off processes which contaminate the working atmosphere
- Analyse area per capita
- Ensure adequate air distribution
- Check for unoccupied heated parts of the building
- Check zonal heating requirements
- Minimise infiltration and uncontrolled air flows
- Measure zonal airchange rates
- Split space-heated areas into zones having different requirements—consider partitioning
- Check that ceiling heights are not excessive—consider the use of false ceilings
- Maximise the extraction of contaminant and minimise the supply and extraction of fresh air
- Ensure that the ventilation system is controlled
- Ensure that opening windows or doors is not used to control temperatures
- Eliminate random infiltration
- Eliminate arbitrary manual adjustment of air flows and temperatures
- Balance inlet and extract fan sets
- Maintain a slight positive pressure to eliminate draughts
- Introduce self-closing exterior doors, plastic or air-curtains, vestibules and air-locks

Energy storage systems

- Check heating arrangements for storage tanks
- Check temperatures of storage vessels
- Check temperature control arrangements
- Check amount and condition of insulation
- Check maintenance procedures
- The use of energy accumulation should be considered to:
  - balance load factors
  - peak lop and use off-peak electricity
  - increase overall energy efficiencies of boilers and distribution systems
  - harness ambient energy

Plant and equipment

- Check conditions of plant and equipment
- Check functions and efficiencies of electrical equipment

- Measure process temperatures and pressures and check if appropriate
- Check efficiencies of plant and equipment
- Check maintenance procedures
- Check control arrangements
- All hardware should be matched to the purpose for which it is required
- All systems should be operated at rates corresponding to maximum efficiency (normally fully loaded in continuous operation)
- Intermittent operations and fluctuations should be avoided
- Efficiency checks should be carried out frequently using standardised procedures
- Plant should be selected on sensible extreme conditions
- Look for heat-recovery opportunities

Lighting

- Check zonal lighting requirements
- Check conditions and cleanliness of luminaires and windows
- Check the maintenance procedures
- Check lighting controls
- Check that parts of the building are not being lit unnecessarily
- Challenge the need for large areas of glazing
- Obtain the economic balance of artificial versus day lighting
- Check colours of room surfaces
- Eliminate glazing
- Keep windows and rooflights clean
- Eliminate luminaires
- Keep luminaires clean
- Replace lamps when their efficiency drops
- Avoid dark background colours
- Use automatic controls
- Zone lighted areas
- Do not light unoccupied areas—use infra-red detecting switches
- Use separate circuits for cleaners and for times outside working hours
- Use separate circuits at the daylighted peripheries
- Never use filament lamps
- Use low-energy fluorescent or discharge lamps
- Maintain lighting systems in good order
- Look for heat-recovery opportunities

Thermal insulation

- Check conditions of buildings, plant and equipment
- Check weather-stripping of external doors and windows

- Check seals on doors leading to stairwells and vertical shafts
- Check self-closures
- Check loading bays
- Check structural integrity
- Seal unused stacks and vents
- Check effective use of building space
- Check thermal insulations of roofs, walls, windows and floors
- Is insulation and draughtproofing adequate and in good repair? Roof, walls, floors, interzones, doors, windows?
- Insulate all high temperature surfaces according to economic optima, bearing in mind maximum refractory temperatures allowable
- The 'law of diminishing returns' applies
- Apply insulation to the inside of intermittently heated enclosures
- Apply insulation to the outside of continuously heated systems
- Conduct a comprehensive transmission heat loss survey
- Specify all insulating opportunities and evaluate cost-effectivenesses
- Produce a schedule describing the optimal sequence for the application of insulating options and compare this with schedules for other energy-saving options
- Evaluate the diseconomies ensuing from reduced firing rates arising from lessening the demand
- Consider the additional diseconomic effects of the likelihood of mechanical damage, moisture or vapour ingress, leading to condensation within the insulant or structure, infestation, fire hazard or deterioration which may occur
- Check vapour barriers and look for interstitial condensation
- Wet insulation acts as a heat pipe thermal fin!
- Insulation also masks problems which may occur in the future
- Incorporate vapour barriers and ensure adequate ventilation to vent off moisture and drainage arrangements should water infiltrate the insulant

Ventilation

- Check air-handling plant
- Check thermal insulation of plant
- Look for blockages
- Check filters for cleanliness
- Check settings and operations of dampers
- Check cleanliness and operations of heating/cooling coils
- Check ventilation rates
- Check ventilation arrangements
- Check how air gets in
- Check how air gets out
- Measure ventilation rates in different zones

- Check air distributions
- Check zonal ventilating requirements
- Check local extract requirements
- Check extract flow rates
- Check operations of fans
- Check if parts of the building are being ventilated unnecessarily
- Check for vertical stratification
- Check maintenance procedures for fans, ducts and filters (sick building syndrome)
- Check control arrangements
- Check for infiltration of outside air at:
  - loading and delivery bays
  - external doors
  - openable windows
  - shafts and flues—consider the introduction of controlled dampers
  - other openings
  - broken glazing
- Consider the introduction of vestibules, air-curtains and the like
- Consider the introduction of automatic door closures
- Minimise infiltration and uncontrolled air flows
- Divide the building into zones having different temperature and airchange requirements—consider the use of partitions and local variable-speed fans
- Check that ceiling heights are not excessive—causing stratification and presenting a large unoccupied volume to heat or cool; consider the use of false ceilings
- Minimise vertical stratification by ensuring adequate mixing throughout the volume
- Consider the use of destratifiers
- Ensure adequate air distribution
- Stick to minimum fresh air requirements—consider the use of obscuration meters; question the minimum airchange rate adopted
- Ensure that the required airchange rate is attained in working areas but do not airchange volumes where fresh air changes are not required (e.g. roof spaces, storage areas, unoccupied rooms or regions)
- When minimum airchange rates are being achieved, consider the introduction of waste heat recovery
- Isolate and vent-off processes which contaminate the working atmosphere
- Maximise the extraction of contaminant and minimise the supply and extraction of fresh air
- Ensure that the ventilation system is controlled
- Check start-up, shut-down and sequencing
- Check controls
- Eliminate random infiltration
- Eliminate arbitrary manual adjustment of air flows and temperatures

- Balance inlet and extract fan sets
- Maintain a slight positive pressure to eliminate draughts
- Look for heat-recovery opportunities

Draughts lead to demands for higher temperatures, which lead to calls for higher ventilation rates to combat excess heating, the opening of windows and doors, activating an insidious spiral towards excess airchanges and room temperatures. Stop this happening!

Air conditioning systems

- Check cooling control arrangement
- Check building/system response to controls
- Check that temperatures and ventilation rates are not excessive
- Check supply and return temperatures
- Check conditions of plant and equipment
- Check that heating and cooling systems cannot conflict
- Specify the exact purpose for which air conditioning is required
- Analyse load profiles
- Do not overcapitalise plant—reduce the demand before increasing the amount of cold air supplied
- Consider ice accumulation to avoid maximum demand changes and on-peak tariffs
- Keep heat-transfer surfaces clean
- Check zonal cooling requirements
- Measure zonal airchange rates
- Split areas into zones having different air conditioning requirements—consider insulated partitioning
- Check insulation levels and conditions throughout
- Check for unoccupied air conditioned parts of the building
- Check maintenance procedures for fans, ducts and filters (sick building syndrome)
- Minimise infiltration and uncontrolled air flows
- Check that ceiling heights are not excessive—consider the use of false ceilings
- Ensure adequate air distribution
- Stick to minimum fresh air requirements—consider the use of obscuration meters; question the minimum airchange rate adopted
- Ensure that the required airchange rate is attained in working areas but do not airchange volumes where fresh airchanges are not required (e.g. roof spaces, storage areas, unoccupied rooms or regions)
- When minimum airchange rates are being achieved, consider the introduction of waste heat recovery
- Isolate and vent-off processes which contaminate the working atmosphere

- Maximise the extraction of contaminant and minimise the supply and extraction of fresh air
- Ensure that the ventilation system is controlled
- Ensure that opening windows or doors is not used to control temperatures
- Eliminate random infiltration
- Eliminate arbitrary manual adjustment of air flows and temperatures
- Balance inlet and extract fan sets
- Maintain a slight positive pressure to eliminate draughts
- Introduce self-closing exterior doors, plastic or air-curtains, vestibules and air-locks
- Look for heat-recovery opportunities

Solar gains, lighting dissipations and high temperature thermal loads, emanating from electronics and electrical systems, should be extracted by cooling windows, louvres, shutters, luminaires or equipment, using air or water at outside environmental temperatures. This avoids the wasteful practice in air conditioning systems of allowing such energy to infiltrate into and so disturb the thermal equilibrium of a room, for which it is necessary to use high grade chilled water or refrigerant to remove the excess heat via a large heat-transfer surface in order to regain comfort conditions.

Domestic hot water systems

- Investigate uses for hot water
- Chronicle water usage patterns
- Check for leaks of hot water
- Check the maintenance procedures
- Check the control arrangements
- Temperatures should be controlled and optimised
- Leaks should be prevented
- Check condition and insulation of hot water storage tanks
- Pipes and storage tanks should be adequately lagged
- Look for heat-recovery opportunities

Compressed air services

- Check condition of plant and equipment
- Check compressor efficiency
- Check the position of the inlet air duct
- Check the maintenance procedures
- Check the control arrangements
- Check the amount of compressed air supplied
- Check delivery temperature and pressure

- Check for leaks
- Prevent leaks
- Check the uses of compressed air
- Check pressures at points of use
- Challenge every use of compressed air—**this is the most expensive energy commodity**

First law of thermodynamics:

Heat = work + change in internal energy
In compression:
Work in = pressure energy + change in internal energy
$m\ c_p\ \Delta T = pV + m\ c_v\ \Delta T$
$m\ c_p\ \Delta T = m\ R\ \Delta T + m\ c_v\ \Delta T$
$c_p = R + c_v$
For air:
$1005 = 287 + 718\ J\ kg^{-1}\ K^{-1}$

Thus 1005 units of work are required to produce 287 units of pressure energy, even with 100% efficient compression. Furthermore, the work (electricity) has been produced in the first place in the conversion of heat to work at 30% efficiency at best.

Thus it requires at least 3350 units of heat to produce 287 units of pressure energy, or 11.7 units of heat to produce 1 unit of pressure energy.

Ensure that minimum pressure is utilised for the required operation. Compressed air is invariably generated at the highest pressure needed in various activities and then throttled down to the pressure of each activity. For example, paint spraying is best accomplished at 40 psi, although air is supplied to adjustable spray guns at 100 psi.

Table 5.27 shows the savings resulting from reduced delivery pressures assuming a compressor efficiency of 75%.

Opportunities should be identified for using the waste heat from the compressed air cooling system, such as for space heating.

- Reduce the generating pressure to a minimum
- Consider interstage bleed-off
- Consider the use of localised booster compressors when higher pressures are unavoidable, especially where usage is intermittent
- Switch off compressors when not in use
- Consider the introduction of compressed air accumulation to:
  - — peak lop to reduce maximum demand payments
  - — use off-peak electricity when possible
  - — balance the load
- Consider the use of pressure-stabilising heated bellows and bags containing phase change (liquid-to-vapour) materials in accumulators.

Table 5.27 Savings by reducing compressed air delivery pressures

Delivery pressure psi(10^5Pa(=Nm^{-2}))	Adiabatic delivery temperature K	Total work done MJ kg^{-1}	Savings by reduction MJ kg^{-1}	% saving	£ saving per £1000 annual bill
100 (6.90)	489	0.20	0	0	0
90 (6.21)	477	0.19	0.01	6	60
80 (5.52)	461	0.18	0.02	10	100
70 (4.83)	443	0.16	0.04	22	220
60 (4.14)	407	0.13	0.07	40	400
50 (3.45)	403	0.12	0.08	42	420
40 (2.76)	377	0.09	0.11	54	540
30 (2.07)	348	0.07	0.13	68	600
20 (1.38)	310	0.03	0.17	87	870
14.7 (1.013)	273	0.00	0.20	100	1000

Notes
For a perfect gas, $P_1V_1^n = P_2V_2^n$ and $P_1V_1/T_1 = P_2V_2/T_2$
Therefore, $T_2/T_1 = (P_2/P_1)^{(n-1)/n} = (V_2/V_1)^{(n-1)}$
For isentropic compression, $n = \gamma$, and then:
$T_2 = T_1\ (P_2/P_1)^{(\gamma-1)/\gamma}$
If $P_1 = 14.7$ psi, $T_1 = 283$ K
If $P_2 = 100$ psi, $T_2 = 489$ K
The work of compression is:
$W = (P_1V_1 - P_2V_2)/(\gamma-1)$
Now, PV = mRT, where R is the characteristic gas constant for dry air = 287 J/kg K and so:
$P_1V_1 = RT_1 = 287 \times 283 = 81221$ J/kg
$P_2V_2 = RT_2 = 287 \times 489 = 140343$ J/kg
$W = -147805$ J/kg = 0.15 MJ/kg
Assuming an efficiency of 75%, the total work supplied is 0.20 MJ/kg.

There are two ways to obtain compressed air—heating at constant volume to increase pressure, and by mechanical compression.

Site air inlets in cool, dry positions—up to 7% of electricity costs can be saved by supplying cold, denser air from outside the building. Often reject air from coolers is recycled through the compressor getting hotter and hotter, lighter and lighter and hence more expensive to compress, as well as worsening cooling and intercooling effectivenesses.

Recover heat from cooling and intercooling systems. Supply outside air for cooling and intercooling systems.

Avoid condensation in pipelines—this must be forced along with the air and wastes pressure energy. Reheat compressed air where possible to increase discharge pressures.

Compressed air should only be used as a last resort. It is not uncommon in manufacturing systems for the cost of compressed air to be of the order of 50% of the entire electricity bill, which itself constitutes the major proportion of the overall energy bill.

Never use compressed air for swarf-blowing and cleaning purposes. Meter compressed air usage. Look for heat-recovery opportunities.

Refrigeration plant and chilled water distribution systems

- Check maintenance and operating procedures
- Evaluate load patterns and operating cycles
- Check conditions of plant and equipment
- Check for evidence of inefficient compressor operation
- Calculate coefficient of performance and energy efficiency
- Check for leaks
- Seal leaks
- Check for leaks of refrigerant or chilled water
- Check the maintenance procedures
- Check operation of condenser fans
- Check cleanliness of air-cooled condenser coils
- Check cooling tower spray water system and water treatment
- Check cooling tower performance
- Check cooling tower outlet-to-inlet by pass air circulation
- Check operations of pumps and valves
- Check the control arrangements
- Check operating pressures and temperatures
- Temperatures and pressures should be controlled and optimised
- Provide monitoring instruments
- Ensure that adequate controls are provided
- Check effective operation of controls
- Check temperature settings
- Check controls for cooling tower and condensers
- Check condition and insulation of cold water storage tanks
- Check condition of insulation and vapour seals on cold lines
- Pipes and storage tanks should be adequately lagged
- Look for heat-recovery opportunities
- Check whether useful heat recovery from the condenser might be accomplished

Steam plant

- Check condition of plant and equipment
- Check efficiency
- Check for steam leaks
- Check for condensate recovery

- Check condition of steam traps
- Check maintenance procedures
- Check control arrangements
- Look for heat-recovery opportunities

Waste heat and materials reclamation

High grade energy (which may be 'hot' or 'cold' with respect to the environmental datum) should not be allowed to be dissipated directly to the environment.

The energy rejected from a high grade process should be collected and re-directed via heat exchangers (or simply fans or pumps) to be employed at another place, collected and stored to be employed at another time, or concentrated for another higher grade purpose using a heat pump or other thermal transformer, as long as these operations are economically justifiable.

Attempts should be made to introduce feedback from energy loss centres to higher grade stations in the energy flow sequence (e.g. by recycling materials, heat pumping or incinerating waste). Attempts to reduce or reuse waste should be made before any attempts at recycling or recovery.

Waste energy and materials should be reused wherever economically possible, ensuring that practical grade, time and space-matched uses have been found for the reclaimed amounts.

- The value of the savings must clearly exceed the cost of recovery
- Before attempting to recover 'waste' heat, ensure that a matched need exists for it
- Attempt to balance the load factor between recovered heat and utilisation
- Evaluate the economics of using flue gas to preheat combustion air
- Consider the use of thermal accumulation at the interface to compensate for phase mismatches
- Consider the grade of energy recovered—match this to the grade of the energy required; the use of heat pumps may be considered
- Evaluate diseconomic effects (i.e. the reduction of plume buoyancy and hence flue gas dispersal effectiveness resulting from recuperating heat from (and hence lowering the temperature of) exhaust gases, and the onset of condensation (especially from sulphurous fuels producing corrosive sulphuric acids) inside the flues)
- Optimise the amount of heat recovered
- Greater energy efficiency always requires an expenditure of materials and vice versa (e.g. the greater the area of a heat exchanger, the more effective the transfer of heat)
- The 'law of diminishing returns' applies
- Consider the direct use of exhaust gases (i.e. for drying or for secondary combustion when the first combustion is reductive)

- Consider the advantages and diseconomies of latent heat recovery from flues.
- Consider recuperative or regenerative heat exchange (with possibly a latent heat exchange facility), heat piping or heat pumping to reclaim waste heat from extract ventilating air in order to preheat fresh air
- Be careful to maintain minimum fresh air requirements if regenerators are adopted

Controls

That which is not measured cannot be controlled.

Energy or materials cannot be readily conserved unless accurate and comprehensive measurements in consistent units are first obtained for all activities within the system boundary.

- Ensure that all sensors are situated in sensible positions
- Use the widest possible bandwidths for heating/cooling systems
- Use the minimum setting for heating systems
- Use the maximum setting for cooling systems
- Control separately areas with different heat (cold) demands (e.g. sundry gains due to solar irradiance, people, lights, equipment, etc.)
- Use optimum start controllers
- Use optimum stop controllers
- Ensure that systems are operated on optimised time schedules
- Switch off equipment when not in use
- Install a computerised *energy management control system* to monitor all electricity use and maximum demand.
- **Remember to track back energy saved to the fuel (money) saved at the boilerhouse**

Best investments

The following lists of optimal energy-saving investments have been extracted from over 100 energy audits.

Summary of low cost options (0 to 1-year payback)

Accounts, records and maintenance
- Management and organisation
- Accountability and responsibility
- Good housekeeping
- Good maintenance
- Repair of leaks

- Metering and records
- Records and accessibility of data
- Education of occupants
- Staff awareness campaigns

Temperatures
- Temperature reductions
- Zonal temperature specifications

Heating
- Avoiding overheating
- Adjusting space heating controls
- Heating period reductions
- Zoning space heating
- Shutting down heat to non-occupied rooms and areas
- Using occupancy sensors
- Using and adjusting auto-off switches
- Locking frost protection thermostats
- Using and setting time clocks—resetting after power cuts
- Using thermostats
- Rationalising heated space
- Installing reflective foil behind radiators
- Fitting night setback controls
- Reducing air infiltration
- Improving thermal insulation
- Improving seasonal boiler efficiency
- Reducing temperature gradients via destratifiers
- Reducing distribution losses via insulation and repairing leaks
- Installing weather compensation controls

Electrical heating
- Elimination of electrical heating
- Use of pay-as-you-use meters

Ventilation
- Optimising temperatures and relative humidities
- Using minimum fresh air requirements
- Resetting ventilation plant time switches
- Installing ventilation plant time switches
- Reducing fan speeds
- Using occupation sensors as controllers for ventilation plant
- Good maintenance
- Switching off fans out of hours
- Using free outside cooling when possible

Boilers

- Modifying stand-by boiler policy
- Using the more efficient boiler as the lead boiler
- Improving boiler control monitoring
- Reducing steam boiler pressure at night
- Good maintenance

Hot water

- Reducing hot water temperatures
- Heating water on night tariffs
- Using timeclocks for water heaters
- Using flow restrictors on domestic hot water taps
- Reducing heat losses via insulation
- Removing redundant pipework

Insulation and weatherstripping

- Insulating boiler plant
- Insulating distribution systems
- Insulating behind radiant panels
- Weatherstripping doors and windows
- Loft insulation
- Fitting automatic door closures

Lighting

- Rearranging light switching
- Eliminating unnecessary lighting
- Illuminance survey to cut or reduce unnecessary lighting
- Reducing illumination levels
- Using lighting controls based upon occupancy sensors
- Using lighting controls based upon master switching
- Using task lighting
- Using lighting switch-off controls using timers
- Decreasing lighting periods in common areas
- Using low energy lamps
- Loss of glazed areas
- Cleaning lights and windows

Refrigeration systems

- Relocating refrigeration equipment away from heat sources
- Timeswitches for chiller plant
- Fitting high efficiency belts and ensuring correct tension
- Good maintenance
- Raising chilled water temperature
- Reducing condenser water temperature

- Reducing scale and fouling
- Reducing auxiliary power requirements—cooling tower fans, and circulating pumps
- Minimising defrost cycles

Fans
- Optimising fan speed
- Good maintenance

Pumps
- Optimising flow rates
- Reducing head losses at bends, elbows, restrictions, orifices, valves, vertical risers, frictional forces, etc.
- Good maintenance

Air compressors
- Compressor servicing and maintenance
- Fitting high efficiency belts and ensuring correct tension
- Optimising intercooling
- Repairing leaks
- Reducing inlet air temperature—ducting outside air to compressors
- Increasing inlet air pressure by cleaning or replacing filters
- Installing maximum demand meters at compressor houses
- Using compressor waste heat to heat domestic hot water or ventilating air
- Paying attention to filters, after coolers, driers, receivers, after heaters and instrumentation

Kitchens
- Using time switches
- Minimisation of oven preheating periods and for the hot storage of food
- Switching off extract fans when not required
- Maximising dishwashing loads
- Using auto-off switches for hot water tanks

Controls
- Turning off equipment when not in use
- Using thermostatic controls
- Optimising thermostat positions
- Paying attention to instrumentation

Electricity
- Eliminating multiple meters to reduce maximum demand charges
- Changing to night tariiffs
- Using cogeneration as a bargaining tool to reduce rates

Steam
- Flash steam recovery
- Good maintenance

Summary of medium cost options (1 to 2-year payback)

Accounts, records and maintenance
- Demolishing certain high energy demand, low-occupancy buldings

Ventilation
- Using occupancy sensors for mechanical ventilation control
- Destratification in full height areas
- Installing two-speed fans in air-conditioning plant

Heating
- Using improved radiator valves
- Using zonal heating controls
- Moving heaters to ground level

Boilers
- Using automatic boiler control
- Using compensators for boilers

Water
- All aspects of water conservation (see Chapter 6)

Hot water (see Chapter 6)
- Installation of showers
- Replacing steam by water heater for domestic hot water
- Installing spray taps for handwashing facilities
- Installing point-of-use water heaters

Insulation and weatherstripping
- Insulating hot oil storage tanks
- Insulating piping, ductwork, flanges and fittings
- Insulating distribution pipework
- Trench sealing
- Loft insulation
- Insulating ceilings
- Insulating calorifiers
- Weatherstripping doors and windows
- Using automatic door closures
- Insulating water heaters and storage tanks
- Using skirts to prevent floor heat losses

Lighting
- Using low energy bulbs
- Using fluorescent lights

Controls
- Optimal heating controls
- Using thermostatic controls
- Using time switches
- Installing optimum start/stop controls

Electricity
- Using variable-speed drives

Electrical maximum demand
- Improving electrical power factors
- Using adequate metering

Summary of high cost options (2-year plus payback)

Ventilation
- Heat recovery from extract air
- Recovery of heat/cold from ventilating air

Heating
- Substituting distributed gas heating

Boilers
- Using compensators on boilers
- Flue gas heat recovery
- Decentralising boiler plant

Hot water
- Heat recovery from waste hot water

Insulation and weatherstripping
- Weatherstripping windows and doors
- Roof insulation
- Building vestibules
- Installing suspended ceilings
- Cavity wall insulation

Refrigeration
- Using variable-speed drives on centrifugal chillers
- Installing a chiller bypass to compensate for outside temperature variations

- Installing heat-recovery systems
- Using absorption chillers for peak lopping electrical maximum demand
- Incorporating ice storage using off-peak electricity rates

Thermal storage
- Introducing thermal storage
- Installing storage heaters to replace electrical heaters

Controls
- Using optimal start/stop controls
- Zonal heating/ventilation control
- Installing building energy management systems

Combined heat and power
- Using combined heat and power

Electricity
- Power factor correction
- Installing high efficiency motors

The **reject side options** are those *end-of-pipe* clean-up processes, such as flue gas desulphurisation, and the use of all the BATNEEC listed in Chapter 2

Energy use improvement project—energy saving options for the site considered

Ventilation losses may be reduced by minimising the rate of fresh air change. Since this is already below the recommended value, it is unlikely that a reduction could be accomplished.

A reduction in mean inside air temperature from 21°C to 18°C would, however, save ventilation losses corresponding to £17 700, which becomes £23 600 of fuel saved at the boilerhouse. This would require close temperature control.

Fabric transmission losses may be reduced by improving insulation or by reducing the mean inside air temperature.

The overall building U-value is dominated by that of the roof, the heat losses through which cost £27 000/annum. The addition of a layer 50–100 mm thick of fibreglass insulant could halve this loss and so save £13 500/annum (£18 000 at the boilerhouse), but this may be an expensive retrofit measure. Quotations from contractors should be requested.

A reduction in mean inside air temperature from 21°C to 18°C would save £10 140 by reduced fabric transmission losses, or £13 520 at the boilerhouse.

Table 5.28 Summary of potential savings

Inside air temperature reduction	£37 120
Roof insulation	£18 000
Boiler control	£5 340
Use of compressor cooling air	£8 447
Abolish electrical heating	£12 571
Reduce electrical supply capacity	£3 000
Total savings	£84 478
Initial energy bill	£244 000
% saving	35
Final energy bill	£159 522

Directly rejected energy accounted for £30 342. A further, more detailed, study of these losses should be made.

Fuel conversion losses may be reduced by proper boiler control, which could increase the annual mean conversion efficiency to 80% and so save £5340/annum on current usage.

The introduction of a condensing boiler could increase overall efficiency by a further 10%, saving £10 680/annum on current usage. Estimates from suppliers should be sought.

Air compressor losses accounted for £20 759 in electricity prices. If the compressor 'cooling' air was redirected to aid in heating the building, £6335 of heating fuel could be saved, resulting in savings of £8447 at the boilerhouse. The uses for compressed air should be further investigated. Each £1 of compressed air saves £3.5 of electricity.

Electrical heating should be abolished and replaced with gas-fired systems, saving £11 717/annum, which becomes £8788 at the boilerhouse due to boiler inefficiency. To this must be added savings on maximum demand premiums of £2783, making a total of £12 571.

All these opportunities for savings are listed in Table 5.28.

All uses for electricity should be investigated and reductions sought. Any resulting reduction in sundry gains must then be made up by the increased (but less expensive) use of gas for heating.

Some of the £36 000/annum for electrical maximum demand and supply capacity charges might also be saved from reductions in compressed air utilisation and other electrical uses.

The heat to power ratio for the site, 2.5, indicates that the installation of a combined heat and power system may be suitable. This should be further investigated, but only after the above less capital-expensive measures have been introduced.

Thus savings in excess of £80 000 are possible for the site. It now remains to obtain estimates for the various retrofit measures and to conduct economic cost-effectiveness calculations.

Table 5.29 Summary of investment opportunities (ranked in order of cost-effectiveness)

Retrofit measure	Annual savings	Capital cost	Straight rate of return %
A Reduce electrical supply capacity	3 000	0	infinity
B Reduce inside air temperature	37 120	1 000	3712
C Abolish electrical heating	12 571	2 000	629
D Boiler control	5 340	2 000	267
E Use of compressor cooling air	8 447	10 000	85
F Roof insulation	18 000	40 000	45
Totals	84 478	55 000	154

Table 5.30 Investment plan (£)

Year	Retrofit measure	Capital cost	Savings in year	Accumulated savings
1	A	0	3 000	3 000
2	B and C	3 000	52 691	52 691
3	D, E and F	52 000	84 478	85 169

Table 5.31 Overall summary

Initial energy bill	£244 000
Gross invested	£55 000
Net invested	zero
Final energy bill	£160 000
Residual capital	£85 169

Tables 5.29, 5.30 and 5.31 summarise the investment opportunities and the resulting project plan.

The introduction of *carbon taxes* would result in further financial gains from this plan.

REFERENCES

(1) *Digest of Environmental Protection and Water Statistics*, UK Statistical Office, Department of the Environment, 1992.

(2) Avallone, E.A. and Baumeister III, T., *Marks' Standard Handbook for Mechanical Engineers*, 9th Edition, McGraw-Hill, New York, 1987.

(3) *U.K. Environmental Protection (Prescribed Processes and Substances) Regulations*, HMSO, 1991.
(4) Eurostat, *Basic Statistics of the (European) Community*, 29th Edition, Luxemburg, 1992.
(5) *Breathing in our Cities*, Parliamentary Office of Science and Technology, House of Commons, 1994.
(6) O'Callaghan, P.W., *Energy Management*, McGraw-Hill, London, 1993.

6

Water, effluents and water management

This chapter deals with water supplies, effluents and water management. The problems arising from increased water usage (e.g. supply shortages, effluent and sewage disposal, pollution and eutrophication of water courses) are discussed and the appropriate legislation relating to water is reviewed. Parameters to consider when assessing water quality are listed and water treatment processes are described. The uses for water are examined and a procedure for water accounting is developed. Thirty-nine steps to optimise water management investments at a site are constructed. Rules for the efficient and cost-effective conservation of water are included, as well as questions for the client interview and a checklist for water managers.

6.1 WATER SUPPLIES

Clean water is essential to plant and animal life. Since the human body is formed of more than 50% water, a normal adult needs to drink about 2.5 litres of fluid each day.

The water cycle

The natural hydrological water cycle relates to the movement of water within the Earth's natural environment due to its release via evaporation, rainfall and the return of water to the oceans.

The oceans cover 71% of the Earth's surface and contain over 95% (1500 million km^3) of the Earth's water. A further 4% (60 million km^3) is locked in the polar ice caps and glaciers and the remaining 1% (15 million km^3) exists as fresh water, mostly ground water, 0.15 million km^3 in lakes and 0.015 million km^3 in rivers. Nevertheless, this fresh water represents 3 million m^3 per head of world population.

The latent heat of vaporisation of water is 2.5MJ kg^{-1} (0.69 kWh kg^{-1}). Solar radiation energy is incident at the outer extremities of the Earth's atmosphere at 1.4 kWm^{-2}. After back-scattering by the atmosphere, the hottest deserts have the capability to evaporate up to 1.4 kgm^{-2} of water in an hour.

Water evaporates to the air from the oceans, lakes and rivers, via evapo-transpiration by plants and perspiration and respiration by animals. It is carried by airstreams as vapour and returns to the Earth's surface as rain, ice, snow or condensation from mists as the conditions for precipitation arise.

Solar radiation evaporates about 85% of the water cycle budget from seas and oceans and 15% from the land, lakes and rivers.

When it falls over land as rain, water soaks into porous rocks and runs off impervious rocks into streams, rivers and lakes. During its journey back to the oceans, water is used by people, animals and plants to sustain life and support the sedimentary cycles.

People have adapted the natural water cycle to satisfy their needs by creating an *artificial water cycle* of collection, storage reservoirs, distribution to consumers, collection of effluents, water treatment works, clean-up and disposal. It has been estimated (1) that about 3% of the water that falls on land is intercepted by people, plants and animals to sustain life.

Chapter 1 showed that *the nitrogen cycle*—the process by which atmospheric nitrogen enters living organisms, being absorbed into green plants in the form of nitrates, the plants being eaten by animals, and the nitrogen being returned to the ecosystem through the animal's excreta or when a plant or animal dies—has now been disturbed by the need to produce ever-increasing supplies of food. Since they cannot be recycled to the atmosphere by the denitrifying bacteria, manmade nitrates eventually end up as run-off via fertilisers, sewage and other waste into lakes, rivers and seas, where *eutrophication* (accelerated ageing by stimulating the growth of oxygen-hungry algae blooms) results.

The UK receives between 600 and 2500 mm of precipitation per annum depending upon location (1). Its drinking water comes from two main sources: surface water (50%) (springs, abstractions from rivers and run-off from upland areas into reservoirs) and underground water-bearing rock strata aquifers (50%) (sandstone, chalk, limestone and gravels).

Abstractions from rivers have a wide variety of uses in industry, power generation and for water treatment works. In the UK, overabstraction from ground and surface waters has led to the drying up of rivers and reduced summer flows. Work by the National Rivers Authority (2) identified 43 sites in England and Wales suffering from serious low flow problems, whilst a total of 92 locations were identified as having low flows due to overabstraction.

The situation has been exacerbated as recent years have seen successive seasons of climatic drought conditions, e.g. in 1991/92 when underground aquifers were not replenished in winter months.

Every person, company or organisation wishing to abstract water from rivers or boreholes must first receive a licence from the NRA, unless the quantity of water extracted does not exceed 5 m^3 per day, or, with the agreement of NRA, does not exceed 20 m^3 per day.

6.2 DISCHARGES AND EFFLUENTS TO WATER

Water pollution

Water can be polluted by (3):

- thermal pollution (especially from power stations), which is dangerous for aquatic creatures
- pathenogenic organisms, creating a public health hazard
- oil pollution
- inert, insoluble material
- readily biodegradable organic material that will result in the depletion or complete removal of dissolved oxygen
- synthetic organic compounds which cause toxicity
- salts of heavy metals which cause toxicity
- eutrophication via nitrate-enhanced algae growth
- acids from effluent discharges or depositions from rainfall
- radioactivity

Water suffers from both natural and artificial pollution. Natural pollution involves animal wastes, leaf-falls, decaying animal flesh, erosion of soil from river banks, run-off of silt, the accumulation of vegetable matter and acid discharges from peat bogs (4). This can cause changes in turbidity, colour, odour, pH levels and oxygen levels and can result in fish mortalities.

Whilst nature provides sinks for much of the pollution created naturally, the pollution added by industrial and other human activity is burdening the natural system.

The main sources of pollution of water courses are leachate from land and landfills, slurries, silage effluent, pesticides, herbicides, nitrates and industrial effluents (e.g. chemicals, metals and solvents). In addition to the nitrogen in sewage, additional nitrates enter water courses from the agricultural use of artificial fertilisers, as well as phosphates and pesticides.

The main types of toxic pollutants that occur are:

- heavy metals (e.g. cadmium, zinc, lead, mercury, copper)
- synthetic organic compounds (e.g. pesticides, solvents, detergents and phenols)
- toxic gases (e.g. chlorine, ammonia)
- toxic anions (e.g. cyanides, sulphide, fluorides)
- acids and alkalis

6.3 WATER AND THE LAW

Chronology: water and effluents

Table 6.1 lists examples of some of the major EC Action Programmes and Directives, UK Acts and Regulations relating to water and effluents in watercourses and sewerage systems. The list is not intended to be exhaustive.

The first Act making it a criminal offence to pollute any British river was the **Rivers Pollution Prevention Act 1876**, which remained in force until 1951. Sewage could only be discharged if the '*best practical and available means*' to render the sewage matter harmless had been used. Industrial effluents were virtually prohibited since it became an offence to allow 'any poisonous, noxious or polluting liquid proceeding from any factory or manufacturing process' to enter any stream.

In 1912, the eighth report of the **Royal Commission on Sewage Disposal (1896–1915)** showed that the 'nuisance producing power of sewage or effluent is broadly proportional to its powers of deoxygenating the water' and proposed a test of the qualities of rivers and sewage effluent measuring the amount of dissolved oxygen taken up over a period of five days (now called the *biological oxygen demand*, or BOD for short).

The **Rivers (Pollution Prevention) Act 1951** placed a general duty on River Boards to maintain or restore the wholesomeness of rivers. These Boards were given the power to issue consents for discharges and to prescribe lawful emissions standards. This Act applied only to inland waters, but was extended to include estuaries and tidal water in the **Clean Rivers (Estuaries and Tidal Waters) Act 1960**.

The Water Act 1973 placed a general duty on the Secretary of State to secure

Table 6.1 EC Action Programmes and Directives, UK Acts and Regulations relating to water

Year	Type	Development
1876	ACT	Rivers Pollution Prevention Act
1906	ACT	Alkali, etc. Works Regulation Act
1937	ACT	Public Health (Drainage of Trade Premises) Act
1945	ACT	Water Act
1951	ACT	Rivers Pollution Prevention Act
1960	ACT	Clean Rivers (Estuaries and Tidal Waters) Act
1961	ACT	Rivers (Prevention of Pollution) Act
1961	ACT	Public Health Act
1971	ACT	Rural Water Supplies and Sewerage Act
1973	ECAP	First EC Environmental Action Programme
1973	ACT	The Water Act

continued

Table 6.1 *(continued)*

Year	Type	Development
1974	ACT	Control of Pollution Act
1974	ACT	Health and Safety at Work, etc. Act
1974	DIR	Quality of Fresh Waters Needing Protection or Improvement in Order to Support Fish Life
1976	DIR	EC Directive on Pollution caused by Certain Dangerous Substances Discharged to the Aquatic Environment of the Community
1976	REG	Control of Pollution (Discharges into Sewers) Regulation
1976	DIR	Quality of Bathing Water
1977	ECAP	Second EC Environmental Action Programme
1979	DIR	Quality Required of Shellfish Waters
1980	DIR	EC Directive Relating to the Quality of Water for Human Consumption
1980	DIR	Protection of Surface and Coastal Waters from Pollution by Dangerous Substances
1981	ECAP	Third EC Environmental Action Programme
1984	REG	Control of Pollution (Consents for Discharges) Regulations
1987	ECAP	Fourth EC Environmental Action Programme
1988	ACT	Environment and Safety Information Act
1989	ACT	The Water Act
1989	REG	Trade Effluents (Prescribed Processes and Substances) Regulations
1989	REG	Drinking Water Inspectorate Water Supply (Water Quality) Regulations
1989	REG	Control of Pollution (Consents for Discharges) Regulations
1989	REG	Control of Pollution (Registers) Regulations
1989	REG	Surface Water (Classification) Regulations
1990	ACT	Planning (Hazardous Substances) Act
1990	ACT	Environmental Protection Act
1991	DIR	EC Directive concerning Urban Waste Water Treatment
1991	REG	Environmental Protection (Prescribed Processes and Substances) Regulations
1991	REG	Environmental Protection (Applications Appeals and Registers) Regulations
1991	REG	Environmental Protection (Authorisation of Processes) Regulations
1991	ACT	Water Resources Act
1991	ACT	Water Industry Act
1991	REG	EC Eco-Audit Regulation
1992	REG	EC Ecolabelling Regulation
1993	ECAP	Fifth EC Environmental Action Programme

Notes
ACT—UK Act
REG—UK or EC Regulation
DIR—EC Directive
ECAP—EC Action Programme

the execution of a national policy for water, including the restoration and maintenance of the wholesomeness of rivers and other inland waters.

The **Control of Pollution Act 1974** required that discharges to rivers, specified underground waters or land must have the consent of the Water Authority.

The EC Directive on the **Quality of Drinking Water for Human Consumption** (80/778/EEC) set standards for drinking water quality to be applied to tap water.

6.4 WATER QUALITY

In classifying water quality, the following variables are considered, covering a wide range of possible pollutants.

Oxygen levels

Biological oxygen demand (BOD)

This is the amount of oxygen required to oxidise the polluting substances in the water. Oxygen used up to react with pollutants leaves less to sustain aquatic life.

Pollutant content

E.g. nitrate, pesticide, herbicide and ammonia content. Table 6.2 lists the properties and measurement units of the various characteristics and contents of water.

Other properties

- *Algae blooms.* Algae are simple plants which thrive in wet environments. If there is an excess of nutrients in the water these algae multiply, profusely forming algae blooms, using up the oxygen in the water and causing eutrophication (5).
- *Alkalinity.* This is the capacity of water to neutralise acids due to its bicarbonate content.
- *Chemical oxygen demand (COD).* This is the amount of oxygen consumed in the complete oxidation of carbonaceous matter in an effluent sample. COD is measured in a standard test which uses potassium dichromate as the oxidizing agent (5). Oxidisability (mg O_2/l) is measured from the *permanganate value.*

Table 6.2 Properties and measurement units of the various characteristics and contents of water

Property	Measurement unit
Colour	mg/l Pt/Co
Turbidity	Formazin turbidity units
Odour	Dilution no.
Taste	Dilution no.
Temperature	°C
Hydrogen ion	pH value
Conductivity	μS/cm
Chloride	Cl mg/l
Calcium	Ca mg/l
Total hardness	Ca mg/l
Alkalinity	HCO_3 mg/l
Sulphate	SO_4 mg/l
Magnesium	Mg mg/l
Sodium	Na mg/l
Potassium	K mg/l
Dry residues	mg/l
Nitrate	NO_3 mg/l
Nitrite	NO_2 mg/l
Ammonium	NH_4 mg/l
Kjeldahl nitrogen	N mg/l
Permanganate value	O_2 mg/l
Total organic carbon	C mg/l
Dissolved or emulsified hydrocarbons	μg/l
Phenols	C_6H_5OH μg/l
Surfactants	μg/l (as laurel sulphate)
Aluminium	Al μg/l
Iron	Fe μg/l
Manganese	Mn μg/l
Copper	Cu μg/l
Zinc	Zn μg/l
Phosphorus	P μg/l
Fluoride	F μg/l
Silver	Ag μg/l
Arsenic	As μg/l
Cadmium	Cd μg/l
Cyanide	CN μg/l
Chromium	Cr μg/l
Mercury	Hg μg/l
Nickel	Ni μg/l
Lead	Pb μg/l
Antimony	Sb μg/l
Selenium	Se μg/l
Boron	B μg/l
Barium	Ba μg/l

continued overleaf

Table 6.2 *(continued)*

Property	Measurement unit
Pesticides and related substances	
1. Individual substances	μg/l
2. Total substances	μg/l
Polycyclic aromatic hydrocarbons (PAH)	μg/l
Tetrachloromethane	μg/l
Trichloroethene	μg/l
Total coliforms	Number/100 ml
Faecal coliforms	Number/100 ml
Organic carbon content (total organic carbon)	mg/l
Dissolved or emulsified hydrocarbons	mg/l

- *Coliforms (faecal coliforms)*. This is a group of bacteria whose absence from drinking water is regarded as a guarantee of freedom from pathogenic (disease-causing) bacteria (5).
- *Conductivity*. This is electrical conductivity, measured in μSiemen/cm
- *DDE*. This is 1,1-dichloro-2.2-bis(*p*-chlorophenyl)ethylene, a metabolite of DDT. It is one of the most abundant organochlorine compounds in the biosphere, being formed by the action of some soil micro-organisms on DDT, and is more inert and persistent than DDT. Sunlight can photodegrade DDE to about nine other products, including several chlorinated biphenyls which can have severe biological effects (5).
- *DDVP*. This is dichlorvos, 2,2-dichlorovinyl, dimethyl phosphate, an insecticide often sold for domestic use (5).
- *pH level*. This is a measure of acidity/alkalinity (for pure water, pH = 7 (neutral)). Carbon dioxide in the atmosphere dissolves in rain, reducing its pH to 5.6. Naturally occurring oxides of sulphur and nitrogen reducc this to 5.0. Lower, more acidic values are produced via the combustion of sulphurous fossil fuels.
- *Surfactants* are surface-active agents.
- *Total dissolved solids (TDS)*. These are the solid residues after evaporating a sample of water or effluent (dry residues).
- *Suspended solids*.
- *Hardness* is a property of water usually manifested as 'needing more soap to get a lather'. It is classed as 'temporary' if it can be removed by boiling the water (depositing a carbonate scale), or 'permanent' if it cannot be removed in this way. Values are expressed as an equivalent amount of calcium carbonate ($CaCO_3$). Water with a hardness of less than 50 ppm of calcium carbonate is classified as being 'soft' (5).
- *Turbidity*. This is reduced transparency caused by scattering of light by suspended particulate matter.

Drinking water quality standards are driven mainly by EC Directives via the **Water Supply (Water Quality) Regulations 1989**. Water sampling is conducted to ensure that the various parameters comply with the *prescribed concentration value (PCV)*.

This was incorporated into UK legislation under **The Water Act 1989**. World Health Organisation (WHO) guideline standards for drinking water also exist.

Not all of the variables are always measured. In Bedfordshire, for example, Anglia Water samples for the following shortened list: colour, turbidity, odour, taste, nitrate level, aluminium, iron, manganese, lead, pesticides and related substances (individual substances, total substances), total coliforms and faecal coliforms.

As can be seen in Table 6.3, there are some differences in particular parameters between WHO (World Health Organisation) and UK/EC values. Differences between UK and EC values have come about because of 'relaxations' in the UK laws allowed by the EC. Those differences between UK/EC and WHO values probably arise because of varying interpretations of what is considered safe.

EC Directives on water quality include a framework Directive on pollution caused by **Certain Dangerous Substances Discharged to the Aquatic Environment** (76/464/EEC) and EC Directive 80/68/EEC, which set out basic rules for the **Protection of Surface and Coastal Waters from Pollution by Dangerous Substances**, listed as:

- List 1: '*Black List*' substances, included for their toxicity, persistence and bioaccumulability—129 substances are listed, e.g. cadmium, certain carcinogens, lindane, mercury, DDT, carbon tetrachloride, pentachlorophenol, aldrin, dieldrin, endrin, chloroform, HCB, HCBD, 1,2-dichloroethane, trichloroethylene and trichlorobenzene. Under no circumstances should List 1 substances be allowed to enter groundwater.
- List 2: '*Grey List*' substances, which are less toxic substances (e.g. chromium, lead, zinc, copper, nickel and arsenic) but discharges of which should be limited to avoid pollution.

Surface waters are protected by EC Directive 75/440/EEC on the **Quality Required of Surface Water Intended for the Abstraction of Drinking Water**. This requires member states to classify the water qualities of surface waters at the point of abstraction into three classes which define the different levels of water treatment necessary. The standards are detailed in the **Surface Water (Classification) Regulations 1989**.

Typical river quality classifications are given in Table 6.4.

The **Royal Commission Report (1912)** classified rivers according to Table 6.5.

EC directives include:

- 74/659/EEC Council Directive on the **Quality of Fresh Waters Needing Protection or Improvement in Order to Support Fish Life**.
- 76/160/EEC Council Directive on the **Quality of Bathing Water**.

Table 6.3 Drinking water quality standards

Parameter	Units	UK	EC	WHO
Colour	mg/l Pt/Co scale	20	20	15
Turbidity	Formazin turbidity units	4	4	5
Odour	Dilution no.	3 at 25°C	3 at 25°C	inoffensive
Taste	Dilution no.	3 at 25°C	3 at 25°C	inoffensive
Temperature	°C	25	25	no set guideline
Hydrogen ion	pH value	5.5–9.5	6.5–8.5	6.5–8.5
Conductivity	μS/cm	1500 at 20°C	400 at 20°C	no set guideline
Chloride	Cl mg/l	400	200	250
Calcium	Ca mg/l	250	100	no set guideline
Total hardness	Ca mg/l	60	60	500
Alkalinity	HCO_3 mg/l	30	30	no set guideline
Sulphate	SO_4 mg/l	250	250	400
Magnesium	Mg mg/l	50	50	100
Sodium	Na mg/l	150	150	200
Potassium	K mg/l	12	12	no set guideline
Dry residues (1)	mg/l	1500	1500	1000
Nitrate	NO_3 mg/l	50	50	10
Nitrite	NO_2 mg/l	0.1	0.1	no set guideline
Ammonia	NH_4 mg/l	0.5	0.5	no set guideline
Kjeldahl nitrogen	N mg/l	1	1	no set guideline
Permanganate value	O_2 mg/l	5	5	no set guideline
Total organic carbon	C mg/l	(2)		
Dissolved or emulsified hydrocarbons	μg/l	10	10	no set guideline
Phenols	C_6H_5OH μg/l	0.5	0.5	no set guideline
Surfactants	μg/l (as lauryl sulphate)	200	200	no set guideline
Aluminium	Al μg/l	200	200	200
Iron	Fe μg/l	200	200	300
Manganese	Mn μg/l	50	50	100
Copper	Cu μg/l	3000	3000	1000

Zinc	Zn μg/l	5000	5000	5000
Phosphorus	P μg/l	2200	5000	no set guideline
Fluoride	F μg/l	1500	1500	1500
Silver	Ag μg/l	10	10	no set guideline
Arsenic	As μg/l	50	50	50
Cadmium	Cd μg/l	5	5	5
Cyanide	CN μg/l	50	50	100
Chromium	Cr μg/l	50	50	50
Mercury	Hg μg/l	1	1	1
Nickel	Ni μg/l	50	50	no set guideline
Lead	Pb μg/l	50	50	50
Antimony	Sb μg/l	10	10	no set guideline
Selenium	Se μg/l	10	10	10
Boron	B μg/l	2000	1000 (3)	no set guideline
Barium	Ba μg/l	1000	100 (3)	no set guideline
Pesticides and related substances				no comparative guidelines
1. Individual substances	μg/l	0.1	0.1	
2. Total substances	μg/l	0.5	0.5	
Polycyclic aromatic hydrocarbons (PAH)	μg/l	0.2	0.2	no comparative guidelines
Tetrachloromethane	μg/l	3	no comparative guidelines	10
Trichloroethene	μg/l	30	no comparative guidelines	30
Total coliforms	Number/100 ml	0	0	0
Faecal coliforms	Number/100 ml	0	0	0

Notes
(1) Dry residues after drying at 180°C.
(2) No significant increase over that normally observed.
(3) These are guideline values only and not maximum limits.

Table 6.4 Typical river quality classifications

Description	Code	Class	Potential use
Good quality	1A	Water of high quality *High amenity value*	Supply abstraction Game reserves High class fisheries
	1B	Water of less high quality than 1A	Supply abstraction Game reserves Fisheries
Fair quality	2	Reasonably good *Moderate amenity value*	Water suitable for potable supply after advanced treatment Coarse fisheries
Poor quality	3	Waters which are polluted to an extent that fish are absent or only sporadically present	Low grade abstraction but considerable potential for further use if cleaned up
Bad quality	4	Water that are grossly polluted and liable to cause nuisance	

Source: *The Bedfordshire Environment, Reports 1 and 2*, Bedfordshire County Council 1991 and 1992.

Table 6.5 Classifications of rivers

Quality	BOD (mg/l)
Very clean	1.0
Clean	2.0
Fairly clean	3.0
Just free of signs of pollution	4.0
Doubtful	5.0
Bad	10.0

- 79/923/EEC Council Directive on the **Quality Required of Shellfish Waters**.
- 91/271/EEC Council Directive concerning **Urban Waste Water Treatment** which sets minimum treatment standards for sewage depending on the nature of the receiving waters and the size of the sewage treatment works.

The **Water Act 1989** introduced a system of statutory water quality objectives and standards for drinking water. It created the **National Rivers Authority (NRA)** as a water pollution 'watchdog', having responsibility for the monitoring and control of water pollution. This Authority controls pollution of rivers, estuaries and bathing waters and manages water resources. It protects against floods, supervises fisheries, nature conservation and recreation for inland waters. The NRA exercises statutory powers for controlling prescribed substances and emissions to water from prescribed processes and audits reports on the activities of **Waste Regulatory Authorities (WRAs)**. Consent from the NRA is required for discharges to inland waters. It became a criminal offence to discharge any poisonous, noxious or polluting matter to any stream or controlled water without this consent. The setting up of the NRA paved the way for the privatised **water companies**, separating the roles of discharger and regulator. These water companies have responsibilities for the qualities of drinking water, rivers and bathing waters. Maximum fines for water pollution offences were increased.

The **Drinking Water Inspectorate Water Supply (Water Quality) Regulations 1989** ensure that water companies fulfil statutory requirements for the supply of quality drinking water.

The **Water Resources Act 1991** introduced substantial provisions for public registers to list consents authorised by the NRA to discharge into controlled waters. The NRA was also given the power to clean up contaminated controlled waters and then recover the costs involved from the polluter. The **Water Industry Act 1991** required that consent be obtained from a water company to discharge trade effluent to sewers.

Thus all liquid emissions are controlled either by the water company or by the National Rivers Authority.

The EC Directive **Concerning Urban Waste Water Treatment 1991** indicated that it would become unlawful to discharge industrial waste into the municipal system without prior authorisation from the relevant authority. It requires that certain standards be applied in terms of treatment in certain industries.

The **Trade Effluents (Prescribed Processes and Substances) Regulations 1989** listed the most dangerous '*Red List*' substances, which became the SCHEDULE 5 substances listed in the **Environmental Protection (Prescribed Processes and Substances) Regulations 1991**. The release into water of these substances must now be authorised by Her Majesty's Inspectorate of Pollution (HMIP) under the requirements of integrated pollution control (IPC).

The **Environmental Protection Act 1990** required that IPC authorisation from HMIP be obtained for those most polluting plants which release the 'Red List' substances into controlled waters or sewers.

Emissions to water

Prescribed substances (SCHEDULE 5) ('Red List')

- Aldrin—agricultural insecticide (carcinogen) (chlorinated hydrocarbon)
- All isomers of DDT (chlorinated hydrocarbon)

- All isomers of hexachlorocyclohexane
- All isomers of trichlorobenzene
- Atrazine
- Azinphos-methyl
- Cadmium and its compounds
- Dichlorvos
- Dieldrin (carcinogen) (chlorinated hydrocarbon)
- Endosulfan
- Endosulphans (chlorinated hydrocarbon)
- Endrin (chlorinated hydrocarbon)
- Fenitrothion
- Hexachlorobenzene
- Hexachlorobutadiane
- Melathion
- Mercury and its compounds
- Pentachlorophenol
- Polychlorinated byphenals (PCBs) (chlorinated hydrocarbon)
- Simazine
- Tributyltin compounds
- Trifluralin
- Triphenyltin compounds
- 1,2-Dichloroethane

Additional substances on priority list

Metals
- Arsenic
- Chromium
- Copper
- Lead
- Nickel
- Zinc

Non-metals
- Azinphos-ethyl
- Carbon tetrachloride
- Chloroform
- Dioxins
- Fenthion
- Parathion
- Parathion-methyl
- Trichloroethane
- Trichloroethylene

Other metals

- Aluminium
- Antimony
- Barium
- Beryllium (extremely dangerous hazardous pollutant)
- Boron
- Calcium
- Cobalt
- Cyanide
- Fluoride
- Hydrogen bromide
- Iron
- Magnesium
- Manganese
- Molybdenum
- Phosphorus
- Plutonium
- Potassium
- Selenium
- Silicon
- Silver
- Sodium
- Tin
- Vanadium

Other non-metals

- Ammonia (NH_3)
- Benzene (C_6H_6) (highly toxic carcinogen)
- Benzene hexachloride
- Bleaching agents
- Chloride
- Chlorine
- Chlorophenols (biocides)
- Detergents
- Dichloroprop
- Dioxins (herbicides, disinfectants, bleaches, burning plastics)
- Fluorides
- Herbicides
- Hydrocarbons (dissolved or emulsified)
- Hydrogen sulphide
- Isoproturon
- Kjeldahl nitrogen (N)
- Lindane

- Methyl butyl amine
- Mercaptans (sulphur compounds produced at sewerage works)
- Nitrate (NO_3)
- Nitrite (NO_2)
- Oils
- Organochlorines
- Organophosphates—azodrin
- Other chlorinated hydrocarbons (insecticides) (major and serious pollutants)
- Particulates
- Permanganate (O_2)
- Pesticides and related substances
- Phenols (aromatic organic compound)
- Phosphorus Oxides (e.g. PO_4)
- Polycyclic aromatic hydrocarbons (PAH)
- Polyvinylchloride (PVC)
- Propyzamide
- Solvents
- Sulphate (SO_4)
- Synthetic pyrethins (insecticides) (major and serious pollutants)
- Tetrachloromethane

The NRA regulates the quality of inland and coastal waters in England and Wales. Its responsibilities include surface and groundwater management, environmental quality, pollution control, flood protection, land drainage, recreation and fisheries. It can prosecute for pollution of rivers.

6.5 WATER TREATMENT

Drinking water

The objectives of water treatment are to produce water with the following characteristics (7):

CLEAR	no turbidity or suspended matter
COLOURLESS	clear and transparent
ODOURLESS	no smell
PALATABLE	no unpleasant taste
SAFE	no disease germs, organisms or harmful mineral content
REASONABLY SOFT	no excessive concentrations of soap-consuming mineral salts

Water can be purified by the natural means of storage and filtration (3). It may be stored in ponds and lakes, where:

- the suspended matter settles to the bottom
- harmful bacteria gradually die out
- colour is reduced by the bleaching effect of the sun
- organic impurities are oxidised in the upper layers

It may be filtered through soil, which removes suspended matter, bacteria and some other impurities by biological action.

Water treatment includes such storage and filtration, as well as screening, aeration, coagulation, flocculation, clarification, mechanical filtration, pH adjustment, disinfection and softening.

Screening

The water is passed through mesh screens before entering the main pumps. These pumps remove water weed and waterborne debris.

Aeration

The water is brought into contact with air by cascading through towers containing a support matrix (i.e. metallurgical coke) or spraying to eliminate odours, to remove objectional dissolved gases, to oxidise metal salts and bring them out of solution into a filterable form.

Coagulation and flocculation

This process is achieved by the addition of a coagulant, such as alum, or a ferrous or ferric salt. This reacts with material in the water to form a floc which settles rapidly, carrying with it finely suspended matter. The floc also entraps bacteria and absorbs colour from the water. Rapid and thorough mixing of the coagulant with the water is achieved in a special mixing channel or flash mixer.

Clarification

Provided that sufficient doses of coagulant are applied, the floc produced in the bottom of the hopper forms into a sludge blanket and the water is clarified.

Mechanical filtration

This removes suspended particles and residual floc and is achieved via the downward passage of water through finely divided inert material, such as sand or anthracite on a support bed of coarser material (usually gravel).

pH adjustment

In the final stage of the water treatment process its acidity is adjusted by the addition of lime. If the water is too acid, it attacks pipes and fittings. If the water is too alkaline, salt deposits may form in the distribution system. The optimal pH level is termed the *saturation pH.*

Disinfection

The final treated water should be potable (safe to drink) and contain no bacteria capable of causing disease. Disinfection is usually achieved by chemical means, such as the addition of chlorine or the use of ozone gas in conjunction with activated carbon (which removes tastes and odours). The use of ultra-violet technology for disinfection is advancing rapidly.

Additives

Sulphur dioxide, which neutralises excess chlorine, and fluoride, which protects against tooth decay, might also be added at the final stage of treatment.

Aluminium may be added in the form of aluminium sulphate, which acts as a coagulant to remove particles that discolour water (5).

Softening

Water hardness is due essentially to dissolved calcium and magnesium salts, which may be removed by precipitation or ion exchange water softening.

Waste water (3)

Once the water has been used, the contaminated waste water is returned via the sewerage system to a waste water treatment plant, and then returned after processing to the watercourses.

The treatment process relies on the combination of physical separation and the augmentation of natural processes to reduce harmful organic and chemical substances. The following stages are involved: screening, grit removal, primary sedimentation, secondary treatment and final sedimentation.

Screening

Untreated sewage first passes through screens to remove rags, plastic, paper, wood and other objects. These 'gross solids' are combusted in incinerators or buried in landfill.

Grit removal

Grit and sand are removed in slowly flowing channels and tanks so that the solids fall out of suspension. This material is sent to landfill.

Primary sedimentation

The sewage is left to stand in sedimentation tanks, where suspended solids fall to the bottom to form a sludge. This sludge is collected for disposal to landfill, use on land as a fertiliser, or often disposal at sea. The disposal of sewage sludge to agricultural land must comply with the regulations outlined in the **Water Act 1989**. HMIP is responsible for monitoring compliance.

The remaining liquor may, at this stage, be clean enough to discharge to a river. If not, it passes on for secondary treatment.

Secondary treatment

Bacteria bed filtration and the activated sludge process both use naturally occurring bacteria to break down organic substances and remove ammonia.

Final sedimentation

At this stage the effluent still contains some solids, but most of the harmful organisms have been removed. The remaining solids are removed in further settling tanks.

The effluent may then be passed into rivers, streams or the sea.

6.6 THE WATER INDUSTRY IN THE UK

Prior to privatisation in 1989, the UK water industry had operated on a regional basis and was publicly owned. Ten water service companies were formed to provide water supplies and sewerage services. A regulatory model was set up to ensure that:

- the government sets standards
- private companies finance and achieve these standards
- public registers police standards

The set government standards are heavily influenced by EC Directives, covering:

- drinking water quality
- disposal of dangerous substances
- quality of bathing water
- disposal of sewage sludge
- urban waste water treatment

The Water Act 1989 provided for the appointment of an independent **Director General of Water Services (OFWAT)**, its function being to consider whether or not the water companies are carrying out their tasks properly and to set maximum charges for water. The **Drinking Water Inspectorate's (DWI)** main task is to ensure that drinking water supplied in England and Wales complies with the standards laid down in the **Water Supply Regulations 1989**. It inspects each company annually and audits the quality of supply water, and can prosecute if water supplied is deemed unfit for consumption.

Her Majesty's Inspectorate of Pollution (HMIP) provides independent advice to the government on pollution control practices (i.e. **BATNEEC**). It issues consents for trade effluent discharges and regulates for discharges of prescribed substances.

6.7 WATER PRICES

Since privatisation took place in the UK, water bills have increased significantly because a major capital investment was required in order to raise the standards of the water industry to those required by the EC.

The price of water and sewerage services and the annual rate of increase grew steadily over the period from 1986 to 1991 (Figure 6.1).

Between 1989 and 1992, utility prices in the UK increased annually as follows:

Water	10.3%
Oil	8.0%
Coal	2.2%
Gas	2.0%
Electricity	0.3%

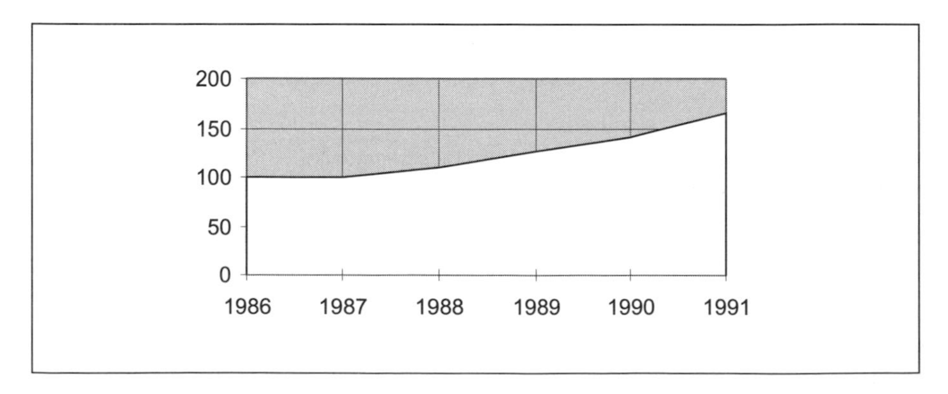

Figure 6.1 Average UK household water bill (index = 100 in 1986)

In 1993, the price for water was £0.48/m^3. It is widely believed that this price might *double* by 1996 and *treble* by the year 2000.

6.8 USES FOR WATER

Demand for water in the UK continues to rise, partly due to increasing population, but mainly due to rising standards of living and the proliferation of uses for water (e.g. for washing machines and dishwashers) and increased frequencies of showers and baths. Over the period from 1970 to 1990, the demand for water in the UK rose from 14×10^9 litres per day to 17×10^9 litres per day and this is expected to rise to over 20×10^9 litres per day by 2020. Industrial demands for water decreased, whilst this decrease was more than made up for by the increasing domestic demand over the period (1). In 1990, 52% of all the water abstracted was used for domestic water supply. The largest sole user of domestic water is for flushing WCs, accounting for some 32%.

Table 6.6 and Figure 6.2 show a typical breakdown of domestic uses for water.

The total volume of water flushed each day down WCs amounts to 4500 million litres, or 12.8% of the total potable water supplied in the UK.

A typical household in the UK consumes about 1 m^3 each day, 356 m^3 per annum, costing £171 in 1993. This might therefore rise to over £500 per annum by the year 2000. This same typical family might use water according to the breakdown given in Table 6.7.

It is interesting to note that, whilst all the water supplied to this household is potable, conforming to EC and UK guidelines, only 0.4% of this water, costing only 70p in a full year, is used for drinking purposes. Considering that a litre of bottled water typically costs £1, the above analysis clearly indicates the extremely good value of tap water in the UK—2000 litres of tap water could be purchased for the price of a single litre of bottled water! It is in order to maintain high quality tap water that the EC Directives and UK standards on drinking water have been formulated.

6.9 WATER MANAGEMENT PROCEDURE

Water management involves the systematic quantification of water-related activities in terms of quantities and qualities of water flowing into the system boundary, through the system and out from the system boundary. The system can comprise a site, a region or a product life cycle.

From an analysis of water bills and water usage patterns, water accounts can be constructed in the same manner as energy accounts. Loss and wastage centres can be identified and rectified. Monitoring and targeting exercises, in comparing

Table 6.6 Domestic uses for water

Activity	%
WC flushing	32
Washing machine	17
Bath/shower	12
Outside use	3
Dishwasher	1
Other miscellaneous uses	35
Drinking	negligible

Source: R. Marchewka, Rain-water Harvesting, MSc thesis, Cranfield Institute of Technology, 1993.

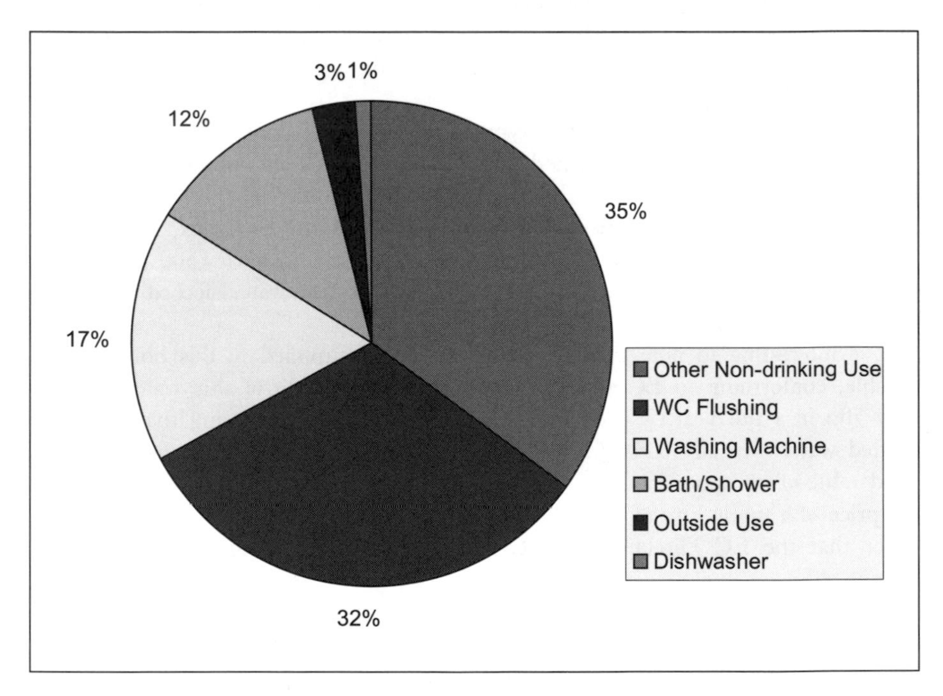

Figure 6.2 Domestic uses for water

current water usage with previous periods, can reveal the onset of new leaks or wastages. Discharge rates of liquid effluent should be correlated with water usage. Possibilities for water cascading, reuse or reclaim should be investigated. Required water qualities for different processes should be specified. Water charges may be based upon rateable value or meter readings. From the water audit, possible financial savings arising from installing a water meter may be estimated.

Table 6.7 How a typical family might use water

Activity	Litres/day	Cost/day (in 1993)	m^3/annum	Cost/annum (in 1993)
WC flushing	300	14.4p	107	£51
Bathing and showering	300	14.4p	107	£51
Dishwashing and cleaning	150	7.2p	54	£26
Laundering	120	5.8p	43	£21
Cooking	96	4.6p	35	£17
Outside use	30	1.5p	11	£5.3
Drinking	4	0.2p	1.5	70p
Total	1000	48.0p	356	£171

Water consumption in like establishments (e.g. schools, hospitals) can be compared by normalising monthly usage by working areas, number of personnel or amount of product. Automatic flushing schedules should be optimised and flushing should not occur when not necessary (i.e. when buildings are unoccupied). Infrared detectors might be employed to detect personnel, or flushings, as well as lights, might be controlled by door openings.

Water accounting

A *water account* of a site is a balance sheet of *water inputs, outputs* and *throughputs* flowing through a site boundary.

The *input* side constitutes an analysis of water bills for a representative annual period.

The *output* side details the quantities and qualities of the ultimate water rejection to the external environment, mainly via leaks to land, evaporation to air, effluent to land, effluent to water courses and effluent to sewers. The data is obtained from a site water survey.

Analyses of *throughputs* may require microaudits, or water balances over individual items of plant and equipment, such as water distribution systems, steam plant, hot water services, processes, washing paraphernalia, WCs and urinals, and effluent management systems.

The fundamental equation for a water audit is as follows:

Water input (grades) + initial impurities (types) + added impurities (types)
= leaks to land (grades)
+ evaporation to air
+ effluent to land (grades)
+ effluent to water courses (grades)
+ effluent to sewers (grades)

The 39 steps to optimise water management investments at a site

1. Identify site and scope of audit.
2. Obtain drainage, effluent and sewage bills for a recent representative year.
3. Obtain water bills for a recent representative year.
4. Examine water supply and effluent capacities, demand variations and bases of charging.
5. Obtain degree-days data for the location.
6. Obtain rainfall data for the location.
7. Identify changes in water usage that have occurred over or since the year under examination.
8. Trace and construct diagram of water handling and distribution system.
9. Trace and construct diagram of surface drainage, sewage and effluent handling, distribution and disposal systems.
10. Estimate number of personnel on site.
    — calculate water/person
    — calculate sewage/person
11. Quantify monthly sewage quantities.
12. Feed data to a spreadsheet.
13. Plot water use versus months of the year.
14. Convert degree-days to mean monthly temperatures during operating hours unless direct temperatures are available.
15. Plot water use versus degree-days.
16. Plot water use versus mean monthly temperatures during operating hours.
17. Correlate with linear regression and quantify scatter.
18. Perform site survey—uses of water and volumes per capita in different uses.
19. Measure water qualities throughout.
20. Quantify different quality water uses:
    — drinking and cooking
    — washing grade 1: human contact (e.g. washbasins, clothes, cutlery and crockery)
    — washing grade 2: non human contact (e.g. vehicle washing, soil watering), flushing of urinals, WCs
    — space heating
    — steam use
    — other processes and appliances
    — other water use
21. Construct output analysis—where does all the water end up?
22. Quantify evaporated water.
23. Look for evaporation prevention possibilities.
24. Investigate and measure effluents—quantities and qualities (solids and dissolved substances).
25. Investigate standards, regulations, Acts and legislation applicable.

26. Identify possible fines and penalties.
27. Identify and cost needs for effluent clean-up.
28. Investigate losses and leaks—quantities and qualities lost.
29. Internal surveys if water-using process plant and equipment significant.
30. Construct throughput analysis.
31. Construct water audit balance sheet—quantities and qualities.
32. Identify options for water conservation and money saving:
    - reduce demand
    - WC flushing (7.5 litres/flush)
    - flush regulators
    - reduction of water pressure (5 litres/minute): tap restrictors
    - control of automatic flushing of urinals
    - elimination or reduction of leaks
    - alternative water supplies
    - recirculation
    - reclamation
    - recycling: effluent clean-up (biological filtration)
    - reusing: (e.g. soil injection)
33. Analyses of options—straight rate of return.
34. Produce ranked investment portfolio.
35. Identify option conflicts.
36. Rerank investment portfolio.
37. Establish five-year investment plan.
38. Set up monitoring and targeting and project management systems.
39. Project manage.

Having constructed the overall water accounts, the various options for system improvement can be identified and evaluated economically. These may be classified into *reject side options* and *demand side options*.

The *reject side options* are those *end-of-pipe* clean-up processes, such as effluent clean-up, waste water treatment and the use of all the BATNEEC listed in Chapter 2.

The *demand side options* include *all the ways in which water may be saved* by employing water conservation techniques and water management.

Rules for the efficient conservation of water

General

- The purposes for which expenditure of water is required should be critically examined.
- As much useful purpose fulfilment should be extracted from a degrading water chain as is compatible with economic and other considerations.
- The quality, not the quantity, of water is the subject of conservation.

- Each water-consuming operation should be examined critically and systematically in isolation and in relation to all other events occurring within the system boundary.
- The manner and extent of all water use should be challenged, including the appropriateness of the process method and the natures of the plant involved.

The following list of questions has been constructed from the perspective of the situation where an independent water auditor is examining a client's site.

Questions for the client interview

Water management procedures

- Who is responsible for water management?
- Position in the organisation? Reporting to?
- Full or part time?
- Qualifications and experience?
- What is done?
- Has a water audit been constructed?
- What has been achieved?

Financial practices

- Who controls the capital spending budget?
- Who controls the recurrent spending budget?
- Upon what financial criteria should cost-effectiveness calculations be performed?
- What is the period available to complete this exercise?
- Is there a list of water-saving investments under review, ranked in order of priority, with detailed costing and cost–benefit calculations?
- If not, why not?

Comments on water consumption

- Is water consumption about right, too high, too low?
- What are the areas of high water consumption?
- How much do you pay for water, sewage and effluent disposal?

Monitoring and recording practices

- How is water consumption reviewed?
- By whom?
- When was the last review?
- How is water consumption analysed?

- Does the analysis normalise the data with level of activity? By building, by product, by month, by year, by cost, by use activity, by sector or section?
- What units of measurement are used?
- What are the metering control arrangements?
- Is there a water consumption forecast/budget?
- Have standards been set? (i.e. for a given task or product or building)
- Is consumption compared with previous periods, other locations, other companies or other industries?
- What are the monitoring and targeting procedures?
- Should a water management policy be implemented?
- What should this policy be?

Personnel water awareness

- Are details of water consumption made known to employees?
- Are employees made aware of the need for water conservation?
- What steps have been made to promote water awareness via education and training, posters, etc.?

Current water conservation measures

- What steps have been taken to reduce water consumption?
- What steps have been taken to cascade or recycle water?

Comments on inefficiencies

- Are there obvious incidents of water wastage?
- Are there leaks of hot water or steam?
- What are the maintenance procedures?
- What are the control arrangements?

Major items

- What are the major water-consuming items of plant and equipment?

Water storage systems

- Is there evidence of leakage?
- What are the maintenance procedures?

Process plant

- Is plant operating efficiently?
- Is there evidence of leakage?

- What are the maintenance procedures?
- What are the control arrangements?

Space heating services

- Is there evidence of leakage?
- What are the maintenance procedures?

Domestic hot water systems

- Are there leaks of hot water?
- What are the maintenance procedures?
- What are the control arrangements?

Chilled water distribution systems

- Is there evidence of leakage?
- What are the maintenance procedures?

Steam plant

- Is plant operating efficiently?
- Are there obvious leakages of steam?
- Is condensate recovered?
- What are the maintenance procedures?
- What are the control arrangements?

Other services

Other plant

Special equipment and processes

Checklists for water managers

General

- Check uses for water
- Is it necessary?
- Is the appropriate quality being used?

Attempts to reduce demand or reuse water should be made before any attempts at recycling or recovery.

Waste water should be reused wherever economically possible, ensuring that practical grade-, time- and space-matched uses have been found for the reclaimed amounts. The value of the savings must clearly exceed the cost of recovery.

Before attempting to recover 'waste' water, ensure that a matched need exists for it. Consider the grade of water recovered—match this to the grade of the water required. Consider the direct use of effluent (i.e. for washing or watering).

Controls

That which is not measured cannot be controlled.

Water cannot readily be conserved unless accurate and comprehensive measurements are first obtained for all activities within the system boundary.

- Ensure that all measurement devices are situated in sensible and accessible positions
- Switch off equipment when not in use

Domestic hot water systems

- Investigate uses for hot water
- Chronicle water usage patterns
- Check for leaks of hot water
- Check the maintenance procedures
- Check the control arrangements
- Leaks should be prevented

Refrigeration plant and chilled water distribution systems

- Check maintenance and operating procedures
- Evaluate load patterns and operating cycles
- Check conditions of plant and equipment
- Check for leaks of refrigerant or chilled water
- Seal leaks
- Check the maintenance procedures
- Check cleanliness of air-cooled condenser coils
- Check cooling tower spray water system and water treatment
- Check cooling tower performance
- Check operations of pumps and valves
- Check the control arrangements
- Check operating pressures and temperatures
- Provide monitoring instruments
- Ensure that adequate controls are provided
- Check effective operation of controls

- Check controls for cooling tower and condensers
- Check condition and insulation of cold water storage tanks
- Check condition of insulation and vapour seals on cold lines
- Lag pipes and storage tanks adequately

Steam plant

- Check conditions of plant and equipment
- Check efficiency of plant and equipment
- Check for condensate recovery
- Check condition of steam traps
- Look for water-recovery opportunities
- Recover as much condensate as possible
- Check pump glands for leakages
- Check boiler feedwater treatment
- Check make-up water quantities
- Check all lines for leaks
- Check for leaks of hot water or steam
- Repair leaks
- Estimate distribution losses
- Check maintenance procedures
- Check control arrangements

6.10 OPTIONS FOR WATER CONSERVATION

Leakage

Leakage is one of the largest water wasters. It has been estimated that 25% of all potable water in the UK is lost through leakage. A single dripping tap can lose 12 litres of water per hour (8).

Water metering

Leaks can be identified by monitoring and targeting flow rates. The installation of a water meter, in place of a nominal charge based upon rateable value, might save money for those households or organisations having lower than average water consumption. Those having higher than average water consumption will inevitably face higher bills. In either case, the existence of a water meter encourages thrift and allows the effects of the introduction of water conservation measures and any subsequent financial savings to be monitored. A water meter is therefore an essential prerequisite for a water management programme, when this is to be implemented in a domestic, commercial or industrial situation. By encouraging thrift, 15% of the annual water bill might be saved.

Taps

Because water delivery pressure is uniformly high, many taps will deliver up to 20 litres of water per minute, whereas less than 15 litres of water per minute is required to prevent splash and 5 litres per minute is sufficient for most purposes.

Aerated taps

These are often used in offices, factories, hospitals and public conveniences and also in some homes. As the water leaves the nozzle it is split up and mixed with air, giving the illusion of a normal water flow but using only 60% of the water which otherwise would be used.

Mixer taps

These are said to save both hot and cold water by allowing faster adjustment to attain a required mixed water temperature. Some modern mixer taps come equipped with infra-red detectors. This detects the presence of hands under the tap's orifice and then runs thermostatically controlled water.

Tap flow restrictors

Kitchen cold taps are connected directly to the incoming mains and may therefore discharge at a rate of 12–30 litres/minute. Bathroom taps, on the other hand, are supplied from a header tank and so discharge at a slower rate of 8–12 litres/minute. For most purposes, these high flow rates are not necessary and much water flows to waste during flow adjustment.

The introduction of flow restrictors reduces flowrates and leads to water savings. Various tap conversion devices on the market include:

- *Pressure compensating valve*, which controls the flow rate in a tap to a pre-selected rate regardless of water pressure. Such a valve may be fitted in the tap or supply pipe.
- *Orifice device*, an orifice plate placed in the flow of the water to restrict its flow rate.
- *Valve stop*, installed inside the tap to limit the amount of opening.
- *Flow reducer*, attached to the tap outflow spout.

- *Sprung taps* are often installed in offices, factories, hospitals and public toilets. These devices are operated by depressing a spring and opening the valve. When the top of the tap is released, the spring slowly returns the valve to the closed position (usually in 5 to 10 seconds).

Showers and baths

The average shower time is of the order of 6 minutes and a typical shower will deliver about 8 litres/minute, using 48 litres of water per shower. This contrasts with the average bath which holds about 80 litres.

A *shower flow restrictor* may be fitted between the taps and the shower hose, or may be incorporated into the shower head.

Toilets/lavatories

A typical water closet uses 8–10 litres per flush and there are normally 8 flushes per day per person, making about 300 litres per day in the average household of four persons. This may be reduced by:

- Inserting a brick or plastic bag into the cistern to reduce the flow quantity.
- Installing a stop on ballcock rise to reduce the flow quantity.
- Installing a two-flush system. (When a short 4 litre flush is necessary, the flush handle is pressed and released. For a long 8 litre flush, the handle is pressed and held down.)
- Installing a low-flush system. (In these 3–4 litre systems, a flush siphon is included below the U-bend. This flush siphon empties to the sewerage system only when full.)
- Installing a mini-flush system. (Water closets which have only 0.8 litres per flush but with 1 bar discharge pressure are available.)
- Replacing the system with an earth closet:
  - the dry closet, which is emptied to a hole in the ground when full
  - the mould closet, which is similar to the dry closet but contains some bacterial mould to speed decomposition
  - the chemical closet, which contains chemicals to kill bacteria and suppress odours (the resulting mixture must be disposed of by specialist effluent disposal contractors)
  - the combusting closet, in which the waste is combusted in a chamber (the residual ash can be used as a fertiliser)
  - the separation closet, in which the waste is separated, the liquid waste draining through a drainpipe directly to fertilise the soil and the solid waste being used as compost

In the UK (after January 1993) new water closets are not allowed to exceed 7.5 litres per flush.

In factories, hospitals, offices and the like, the flushing of urinals is often the greatest single water consumer. Automatic flushing of urinals may be operated in conjunction with infra-red detectors. Companies supplying such systems claim up to 75% water savings using this device.

Dishwashers and washing machines

Currently about 10% of homes in the UK have a dishwasher, whilst 80% have washing machines. The average washing machine uses 100–120 litres of water per full wash, although new systems on the market can work with 70–80 litres per wash. Washing machines should be operated only with full loads.

The *reject side options* are those *end-of-pipe* clean-up processes, such as effluent clean-up, waste water treatment and the use of all the BATNEEC listed in Chapter 2.

A summary of water conservation measures and typical savings arising are provided in Table 6.8.

Rainwater harvesting

Rain water collected from a roof area may be used for flushing the toilet, to water the garden and for washing cars. In order to match supply and demand, a water storage system is required, and it is the size of the storage tank that dictates the economics. The annual precipitation for England and Wales is of the order of 1000 mm per annum, and up to half of this will evaporate back into the air, the remainder being absorbed by soil or drained to watercourses or sewers. Rain water, although sometimes acidic from 'acid rain' content, has virtually no bacterial content before it touches the ground. Bacteria may be picked up, however, from roofs and other collecting surfaces. Thus disinfectants (i.e. chlorine, chloromine, chlorine dioxide, sodium hypochlorate or ozone) may be added to combat these. Existing rain water harvesting systems often incorporate graded gravel filters, placed before the store container, alkaline additives to neutralise acidity levels and fine filters, placed between the store container and the toilet cistern or outside tap (4). The water in the store should also be protected from the light to inhibit algae growth.

Grey water recycling

The reuse or cascading of water has its largest potential in industry, for cooling and process needs. Grey water may also be used for recharging groundwater sources, where natural biodegradation and filtration will ensue, or for subpotable uses, such

Table 6.8 Summary of water conservation measures and typical savings arising

Option	Reduction in consumption	Typical annual water savings (litres (% of total))[1]	Typical payback period (years)[2]
Water supply			
Water metering	15%	40 000 (16%)	5–12
Mixer taps	20%	15 000 (6%)	14–16
Mixer taps with flow level valve	30–60%	30 000 (12%)	6–8
Mixer taps with flow reducer	30–60%	36 000 (14%)	10–12
Mixer taps with infra-red detector for on/off	75%	56 000 (22%)	1–2
Tap valve stops	30%	15 000 (6%)	negligible
Tap flow reducers	20%	15 000 (6%)	2
Sprung taps	40%	30 000 (12%)	not applicable for a home
Showers			
Shower stop	20–25%	15 000 (6%)	1
Flow reducer	25–30%	18 000 (7%)	0.9
Shower head flow reducer	40–50%	30 000 (12%)	1
Shower ball flow reducer	40–50%	30 000 (12%)	0.75
Shower with infra-red detector	75%	50 000 (20%)	1–2
Bath			
Combined bath-tub and shower	35–45%	72 000 (28%)	2–3
Water closets			
Brick in cistern	20%	11 000 (4%)	negligible
Water-filled plastic bag in cistern	20%	11 000 (4%)	negligible
Flow weight stop	50%	30 000 (12%)	0.75
Two-flush system	40–50%	30 000 (12%)	0.6–6
Low flush system	65%	50 000 (20%)	7
Mini-flush system	90%	70 000 (27%)	7
Earth closet, etc.	100%	78 000 (30%)	0.5–2
Urinals			
Flush with infra-red detector	75%		not applicable for a home
Washing machines			
New water-efficient machine	20–30%	16 000 (6%)	40–50
Rainwater harvesting		50 000 (20%)	30
Grey water flushing		77 000 (30%)	38

Notes
(1) For a family with four people, each using 175 litres/day (255 500 litres per annum).
(2) At 1993 device prices and water costs.

Source: F. Frengler, *Water Management*, Cranfield Institute of Technology, 1993.

as recreational lakes and aquaculture. Disinfected 'used' water can also be used for agricultural purposes.

Waste water from lavatories and sinks is termed 'black' water waste, whilst that from baths and showers is termed '*grey*' water waste. Grey water recycling systems recycle the grey water waste to flush toilets. Again, a storage tank is required to match supply and demand. Extreme care must be taken to prevent this grey water contaminating the drinking water system. Existing grey water recycling systems often incorporate grease separation tanks and graded gravel filters (4). Disinfectants are often added to the water in the storage tank to combat bacteria growth. Whilst soapy grey water should not harm garden plants, water contaminated with detergents from washing machines might be harmful, and so it is not recommended that waste grey water be employed for garden watering if this is likely to be the case.

The level of public health risk associated with the use of reclaimed waste water is related to the extent of human contact and the overall level and reliability of treatment provided. Contaminants of concern to public health found in reclaimed waste water are broadly classified as either pathogens or organic compounds. If the level of contact with humans is likely to be high, the protection of public health dictates that pathogenic agents be eliminated from the waste water by disinfection prior to use.

6.11 IMPROVEMENT PROJECT

Having assessed, evaluated and ranked the various retrofit water-saving options in order of cost-effectiveness, the *water* improvement projects can be interleaved into the *energy* **investment portfolio**, the **project plan** and the **project management** and **monitoring and targeting programmes**.

REFERENCES

(1) R. Marchewka, Rain-water Harvesting, MSc thesis, Cranfield Institute of Technology, 1993.
(2) Friends of the Earth, *Draining Our Rivers Dry*, Briefing Sheet, August 1992.
(3) P. Kenny, Water Conservation and Auditing, MSc thesis, Cranfield Institute of Technology, 1992.
(4) F. Frengler, Water Management, Cranfield Institute of Technology, 1993.
(5) A. Porteous, *Dictionary of Environmental Science and Technology*, Open University Press, Buckingham, 1991.
(6) *The Bedfordshire Environment, Reports 1 and 2*, Bedfordshire County Council, 1991, 1992.
(7) Severn Trent Water, *Water Supply*, Severn Trent Information Services, January 1990.
(8) Water Services Association, *Water Wise – Household Guide to Water*, Water Information, London, December 1991.

7

Materials and waste management

This chapter deals with waste management. The problems of resource depletion and pollution arising from consumerism, the throwaway society and increased rates of waste generation are reviewed (e.g. pollution of soil, water and air by solid, liquid and gaseous wastes). Appropriate legislation and regulations are listed. Waste arisings are classified into inert wastes, semi-inert wastes, putrescible wastes, difficult wastes, special wastes and prohibited wastes. The various methods of waste disposal are evaluated and compared. The procedures involved in materials auditing, as well as rules for the efficient conservation of materials, are developed. Questions for the client interview and checklists for waste managers are listed. Thirty-nine steps to optimise waste management investments at a site are constructed. Techniques of materials conservation (e.g. reclaim, recycle, reuse, energy from waste, waste-derived fuels) leading to cost-effective waste minimisation are reviewed and evaluated.

7.1 MATERIALS, PRODUCTS AND POLLUTION

Today's western society is built on consumerism. The more products are produced and sold, the more employment, the more wages, the more spending power, the more products are bought, the more products are again produced, the more products are discarded, the more waste is dumped to land and the greater the pollution of the environment.

In the UK, USA and Europe, well over 33% of all energy and materials are consumed in road and rail building, bridge construction, maintenance, vehicle manufacture, garages, vehicle repairs, petrol and diesel fuels. More roads tend to lead to more traffic, declining public transport, more roads, more traffic, distributed manufacturing sites, more traffic, more roads . . . 60% of all air pollution emanates from motor vehicles, which themselves eventually end up in the scrapyard as waste to land.

In the manufacture of consumer products, energy, water and materials are consumed and pollution of air, water and land by gaseous, liquid and solid wastes results. Product manufacture involves the exploration, extraction, supply, conversion and utilisation of fossil fuels and raw materials in the manufacture, advertising, marketing, distribution, retailing and utilisation and the ultimate disposal of the products. Conservation conscious planning and design can reduce energy and materials consumption and the concomitant environmental pollution in each of these activities.

Environmental auditing should adopt a 'cradle-to-grave' approach. Environmental ratings and comparisons among products should consider the energy and raw materials requirements and the air, water and soil pollution produced in production, distribution, utilisation and ultimate disposal of a product to the environment. The environmental impact of a product should begin at the design stage, commencing with the question: *Is this product really necessary?* 'Green' consumerism, acting locally in response to perceived global problems, may one day reject items considered to be unnecessary or undesirable.

To ensure sustainability, products should be **built to last, built simple and built modular** (1). Unfortunately, to sustain the consumer cycle, the most successful products are those which are designed to be thrown away or rejected to the environment as soon as possible after purchase, such as packaging, newspapers and journals, junk mail, body sprays, hairsprays, shower gels, disposable razors, kitchen rolls, air fresheners, disposable plates, cups and cutlery, batteries, etc.

Toxic materials, unnecessary packaging, disposable or unnecessary products, dangerous products, irreparable items, energy-intensive products and systems, and environmentally polluting items (in manufacture, use or disposal) should be avoided.

Some of the various items entering these categories include clingfilm, aluminium foil, food processors, electric carving knifes, electric can openers, toasted sandwich makers, coffee machines, cosmetics, shower gels, bath foams and oils, hair gels and sprays, home perm kits, plastic furniture, foam upholstery, dishwashers, detergents and washing powders, bleaches, dry cleaning fluids, cleaning liquids, fly and wasp killers, air fresheners, toilet blocks, descalers, toilet cleaners, bath foams, power tools, batteries, electrical lawn mowers, strimmers, shredders, etc., synthetic fertilisers, artificial insecticides, weed controllers, insect repellents, etc., etc,—the list is endless!

The 'throwaway' society

Over 15% by weight and 25% by value of all products purchased in the UK are designed to be thrown away almost immediately; 25% of the plastic produced and 50% of the paper produced is used for packaging, newspapers and plastic cups and cutlery (2).

Each year, UK consumers produce 23 million tonnes of domestic rubbish, 100 million tonnes of industrial waste, 130 million tonnes of mineral spoil, 200 million tonnes of agricultural waste and 30 million tonnes of sewage. It costs £650 million to collect, transport and dispose of these discarded materials.

Some 60% of domestic refuse might be recoverable, worth £2 billion/annum. This would involve waste sorting and collection, shredding organic materials to produce compost, recycling paper, glass, cans and waste oils. Metals, toxic chemicals, paint, sump oil, organic materials, bottles and paper should be separated out prior to disposal.

It is much more cost-effective to reuse a product than to recycle its constituent materials. Almost every discarded material can be reclaimed, reused, recycled or combusted to produce heat, work, fuels, chemicals or materials. There exist no technical barriers to waste recovery, but the recovery of energy and materials requires the expenditure of energy and materials. Thus recovery projects, albeit stimulated by concerns for environmental conservation, are also motivated by economics and social factors.

7.2 WASTE AND THE LAW

Chronology: waste and land pollution

Table 7.1 lists examples of some of the major EC Action Programmes and Directives, UK Acts and Regulations relating to waste and land pollution. The list is not intended to be exhaustive.

Release into land

Prescribed substances (SCHEDULE 6)

- *Alkali metals and their oxides and alkaline earth metals and their oxides.* Alkalis have pH values equal to or above 8. They have the power to neutralise acids and form salts.
- *Azides.*
- *Halogens and their covalent compounds*, e.g. fluorine, chlorine, iodine, bromine and astatine. Halogenated fluorocarbons is the group name for ethane- or methane-based compounds in which some or all of the hydrogen in their structures is replaced by chlorine, bromine and or fluorine. The chlorofluorocarbons (CFCs) are a major part of this family (3).
- *Metal carbonyls.*
- *Organic solvents*, e.g. acetic acid, acetone, benzene, cyclohexanol, ethanol, ethylene glycol, furfural, glycerol, hexane, isopropanol, methanol, methylethylketone (MEK), n-propanol, toluene, trichloroethylene, trichloroethane, white spirit. All can render ground water unpotable on grounds of taste (3).
- *Organo-metallic compounds.* Organophosphates are a group of chemical pesticides, such as azodrin, malathion, parathion, diazinon, trithio and phosdrin,

Table 7.1 EC Action Programmes and Directives, UK Acts and Regulations relating to solid waste

Year	Type	Development
1906	ACT	Alkali, etc. Works Regulation Act
1936	ACT	Public Health Act
1947	ACT	Town and Country Planning Act
1961	ACT	Public Health Act
1972	ACT	Local Government Act
1973	ECAP	First EC Environmental Action Programme
1974	ACT	Control of Pollution Act
1974	ACT	Health and Safety at Work, etc. Act
1975	DIR	EC Framework Directive on Waste
1977	ECAP	Second EC Environmental Action Programme
1978	DIR	EC Toxic and Dangerous Waste Directive
1978	ACT	Refuse Disposal (Amenity) Act
1980	REG	Control of Pollution (Special Waste) Regulations
1981	ECAP	Third EC Environmental Action Programme
1985	ACT	Food and Environmental Protection Act
1986	ACT	The Single European Act
1987	ECAP	Fourth EC Environmental Action Programme
1988	REG	Collection and Disposal of Waste Regulations
1988	ACT	Environment and Safety Information Act
1989	ACT	Control of Pollution (Amendment) Act
1989	REG	Trade Effluents (Prescribed Processes and Substances) Regulations
1989	REG	Control of Pollution (Registers) Regulations
1990	ACT	Planning (Hazardous Substances) Act
1990	ACT	Environmental Protection Act
1991	DIR	EC Framework Directive on Waste
1991	REG	The Environmental Protection (Prescribed Processes and Substances) Regulations
1991	REG	Environmental Protection (Applications Appeals and Registers) Regulations
1991	REG	Environmental Protection (Authorisation of Processes) Regulations
1991	REG	EC Eco-Audit Regulation
1992	REG	EC Ecolabelling Regulation
1993	ECAP	Fifth EC Environmental Action Programme

Notes
ACT—UK Act
REG—UK or EC Regulation
DIR—EC Directive
ECAP—EC Action Programme

which are all descendants of World War II nerve gases and block the central nervous system. The result is hyperactivity and then death (3).

- *Oxidising agents*. These are chemicals which either give out oxygen in chemical reactions or supply an equivalent element such as chlorine to combine with a reducing agent. Atmospheric oxidising agents include ozone and nitrogen dioxide.
- *Pesticides*, e.g. Aldrin, Dieldrin, Endrin, Isodrin. *The Green Consumer Guide* (2) lists 43 garden chemicals used for insecticides, herbicides or fungicides that are dangerous to animals, birds, fish, bees and other insects, irritants and possible carcinogens.
- *Phosphorus*. A toxic chemical element which will cause burns if it touches the skin and is poisonous if swallowed.
- *Polychlorinated dibenzo-p-dioxin* and any congener thereof.
- *Polychlorinated dibenzofuran* and any congener thereof.
- *Polyhalogenated biphenyls, terphenyls and naphthalenes*. Polychlorinated biphenols (PCBs) are stable compounds, which were used extensively in electrical fittings and paints. Although they are no longer manufactured, they are extremely persistent and remain in huge quantities in the atmosphere and in landfill sites. They are not water soluble and float on the surface of water, where they are eaten by aquatic animals and so enter the food chain. They are fat-soluble and so are easy to take into the system, but difficult to excrete (4).

Other damaging non-metals which are not listed as prescribed substances for release to land

- *Asbestos (magnesium silicate ($MgSiO_4$)) materials*, a fibrous mineral substance used as a shield against fire and as an insulating material in loose fibre form. Asbestosis is a disease of the lungs caused by inhaling asbestos dust.
- *Bleaching agents*. These burn on contact with the skin and embrittle organic fibres.
- *Fluorides*, any compound of fluorine, usually found with sodium, potassium or tin. Sodium fluoride is added to drinking water to prevent tooth decay, but can result in deformed growing teeth.
- *Leachates*, substances which are washed out of the soil, such as excess chemical fertilisers, nitrates, phosphates and bromides. They form at the bottom of landfill and, if not contained, contaminate land, watercourses and the water supply.
- *Mercaptans*, extremely noxious sulphur compounds produced at sewage works.
- *Plastics*. These produce poisonous gases when burnt.
- *Polyvinylchloride (PVC)*. When burnt, PVC releases dioxins. PVC powder causes pneumoconiosis.
- *Chlorinated hydrocarbons (organochlorides)*. These are major and serious carcinogenic pollutants used as insecticides. They include DDT, Aldrin, Benzene

Hexachloride, Dieldrin, Aldrin and Lindane. They are very persistent, with a half-life of up to 15 years. They are most harmful to animals such as bees and fish. They enter the food chain and kill small animals and birds that feed on insects, which often become immune to the insecticide (3).

- *Arsenic*. This is a chemical element which forms poisonous compounds, such as arsenic trioxide, and is used to kill rodents.
- *Barium*. This is a chemical element forming poisonous compounds and is used as a contrast when taking X-ray photographs of soft tissue such as duodenal ulcers.
- *Boron*. This is the element from which the control rods in nuclear reactors are made.
- *Cyanide*. This is any salt of hydrocyanic acid. It is present in many effluents from industrial processes. Hydrogen cyanide is a gas that kills very rapidly when drunk or inhaled (3).
- *Herbicides*. 2,4-dichlorophenoxyacetic acid causes lymph cancer (2).
- *Selenium*. This is a non-metallic trace element which is a systemic poison used in the electronics industries and is present in paints and rubber compounds.

Metals

All metals are toxic. The most common metals include:

- *Aluminium*.
- *Antimony*.
- *Beryllium*. This is a metal used in making various special alloys, particularly in aerospace industries. It is an extremely dangerous hazardous pollutant. Berylliosis is poisoning caused by breathing in particles of beryllium. Its poison can even be absorbed by touch.
- *Cadmium*. Cadmium is also used for making the control rods in nuclear reactors. It is present in tobacco smoke and is found in fish and shellfish such as oysters (4).
- *Calcium*. This is a metallic chemical element which is naturally present in bones, teeth and limestone and chalk (calcium carbonate) from dead animals. Quicklime (calcium oxide) is used for industrial purposes and is added to soil to neutralise acidity. It occurs in milk, cheese, eggs, eggshells and certain vegetables. ‘Hard’ water is produced when water passes through limestone, absorbing calcium.
- *Chromium*. This is a metallic trace element used to make alloys for shiny car bumpers and the like.
- *Copper*.
- *Iron*.
- *Lead*. This is a very dense soft metallic element, which is poisonous in compounds and is used as a shield in nuclear reactors. In leaded petrol, lead

tetraethyl is added as an anti-knocking agent and to improve combustion performance. Lead poisoning is caused by imbibing or absorbing lead salts, affecting brain development. Lead can enter the diet through drinking water that has flowed through lead piping, from lead-coated toys (toy soldiers) and from leaded paints.

- *Magnesium.* This is a white metal used in alloys. It is highly combustible.
- *Manganese.* This is a trace element used in steel making.
- *Mercury.* This is a poisonous liquid metal. Its vapour coats the lungs, forming a barrier to the passage of oxygen and preventing oxygenation of the blood.
- *Nickel.* This is a metallic element commonly used in computer wiring.
- *Plutonium.* This is a naturally occurring radioactive element also made in nuclear reactors. It is extremely poisonous and carcinogenic.
- *Potassium.* This is a metallic element which is combustible in contact with water.
- *Silver.*
- *Sodium.*
- *Vanadium.*
- *Zinc.*

7.3 WASTE CATEGORIES

Waste may be classified into:

- inert wastes
- semi-inert wastes
- putrescible wastes
- difficult wastes
- special wastes
- prohibited wastes

as follows.

Inert wastes

These materials do not decompose.

Examples are subsoil, topsoil, hardcore, brickwork, stone, concrete, clay, clinker, coal, coke, sand, silica, cement, ash, plastics, glass, pottery, china, enamels, ceramics, mica and abrasives.

All may be used for aggregrate, hardcore or land infill. Coal and coke can be combusted if they can be separated out.

Semi-inert wastes

These materials decompose slowly and are slightly soluble in water.

Non-combustible materials

For example, plaster, plasterboard, builder's rubble, excavated road metal, metals (iron, steel, aluminium, brass, copper, tin, zinc, gold), boiler scale, diatomaceous earth, oxides of iron, magnesium, zinc, aluminium, copper, titanium, hydroxides of iron, calcium carbonate, calcium sulphate, magnesium sulphate.

The metals are valuable if they can be recovered.

Combustible materials

For example, wood, wood products (hardboard, chipboard, etc.), leather, wool, cotton, linen, hemp, sisal, hessian, string, rope or any other manmade or natural fibre, cork, ebonite, kapok, trees, bushes, garden and horticultural waste, carbon.

All these organic materials can be combusted to provide heat.

Putrescible wastes

These may decompose and may consist in part of soluble matter which could cause pollution if allowed to enter ground or surface water systems.

Non-combustible materials

For example, shotblasting residues, slag and pulverised fuel ash (PFA), foundry and moulding sand, empty metal and glass containers, machinery.

Combustible materials

For example, paper, cardboard and fibreboard, empty paper containers, sacks.

Combustible but heavily polluting materials

For example, empty plastic containers, floor sweepings, electrical fittings, rubber. These should not be combusted but bagged and handed to a waste disposal contractor for safe disposal.

Food waste

For example, waste food and food processing materials, vegetable matter. These may be processed in various ways to produce solid, liquid or gaseous fuels, animal feed or compost.

Other wastes

For example, calcium chloride, sodium chloride, soap and other stearates, cosmetic products, household wastes (or similar wastes from trade, commerce or industry), animal carcasses or parts thereof, aqueous wastes which contain dissolved or suspended inert, semi-inert or putrescible materials, e.g. liquid cosmetic products.

Difficult wastes

These are materials which are not 'special wastes' within the meaning of the ***Control of Pollution (Special Waste) Regulations, 1980****, but which are, or could be, difficult or dangerous to handle or dispose of.*

They may consist, in part, of soluble material which could cause pollution if allowed to enter ground or surface water systems. They consist of materials as follows:

Inorganic acids

- hydrochloric acid
- sulphuric acid
- nitric acid
- chromic acid
- phosphoric acid
- hydrofluoric acid

Organic acids and related compounds

- aliphatic acids, e.g. formic, acetic, oxalic
- aromatic acids, e.g. benzoic, phthalic
- acid anhydrides, e.g. acetic, phthalic
- acid chlorides, e.g. acetyl, benzoyl
- sulphonic acids

Alkalis

- hydroxides of sodium, potassium or calcium
- oxides of sodium, potassium or calcium
- carbonates of sodium or potassium
- proprietary alkaline cleaners

Non-toxic metal compounds

- iron compounds
- titanium compounds
- trivalent chromium compounds

Polymeric materials and precursors

- epoxy resins
- polyester resins
- phenol-formaldehyde resins
- polyurethane
- other resins and polymeric materials
- latex, latex and rubber suspensions and solutions
- synthetic adhesive wastes
- ion exchange resins

Miscellaneous

- dried or denatured paint, varnish or lacquer
- tar, pitch, bitumen or asphalt
- cellulose wastes
- tannery sludge
- oil interceptor wastes

Special wastes

These are materials which could subject persons or animals to risk of death, injury or impairment of health, or could threaten to pollute or contaminate any water supply.

They include the following:

- antimony and antimony compounds
- arsenic and arsenic compounds
- barium and barium compounds
- boron and boron compounds
- cadmium and cadmium compounds
- copper compounds (except oxides)
- chromium (hexavalent compounds)
- lead and lead compounds
- mercury and mercury compounds
- nickel and nickel compounds
- phosphorus and phosphorus compounds

- selenium and selenium compounds
- silver and silver compounds
- tellurium and tellurium compounds
- thallium and thallium compounds
- vanadium and vanadium compounds
- zinc compounds (except oxides)
- biocides and phytopharmaceutical compounds
- laboratory chemicals
- pharmaceutical and veterinary compounds
- tarry materials from refining and tar residues from distilling
- heterocyclic organic compounds containing oxygen, nitrogen and sulphur compounds
- inorganic cyanides
- inorganic halogen-containing compounds
- inorganic sulphur-containing compounds
- organic halogen compounds (excluding inert polymeric materials)
- any medicinal product available on prescription only
- any material with a flash point of 21°C or less
- beryllium and its compounds
- radioactive waste as qualified by the **Control of Pollution (Special Waste) Regulations, 1980**
- all chemical forms of asbestos, either dry or wet, including the following: fine dust, loose fibre, sweepings, offcuts, broken pieces, sacks and bags, sludges and slurries containing:
  - — any measurable quantity of crocidolite (blue asbestos)
  - — chrysolite (white) or amosite (brown) in excess of 1% by weight of free fibre or dust

'Hard wastes' containing bonded asbestos such as asbestos cement, jointing or other asbestos having bitumastic, cement, plastic, resinous or rubber binders are not regarded as 'special wastes' unless they arise in the form of dust from machining or size reduction operations. Such blended asbestos products not in dust or fibrous form may be considered as being semi-inert wastes.

Prohibited wastes

- Peroxides, chlorates, perchlorates, azides and all prescribed substances.
- Explosive materials.
- Radioactive materials as qualified by the **Radioactive Substances Act, 1960** other than those qualified by the **Control of Pollution (Special Waste) Regulations, 1980.**

Alternative waste definitions are provided by Croner (5) as follows:

- *controlled waste*—household, commercial and industrial waste
- *municipal waste*—waste that a Waste Collection Authority (WCA) has a duty to collect
- *civic amenity waste*—wastes that are derived from the household and which Waste Disposal Authorities (WDAs) have a duty to receive at collection points, not arising from business, commerce or industry
- *notifiable waste*—wastes that are poisonous, noxious or polluting

7.5 WASTE ARISINGS

According to the UK Department of the Environment (6) the estimated total annual waste arisings in the UK is of the order of 400 million tonnes, distributed by sector as indicated in Table 7.2.

The average European citizen discards about 1 kg per day of domestic waste, having a mean density of about 130 kg/m^3. A typical analysis of municipal refuse in a northern European city may contain constituents as indicated in Table 7.3

7.6 MATERIALS USES AND OPTIONS FOR IMPROVEMENT

Materials and waste accounting

Within an organisation, an *input/output/throughput* analysis of all materials flows, other than fuels and water, is exceedingly tedious to perform. A thorough examination of purchase orders and inventories is required to construct the *input* database. The uses of all materials should be identified. All these materials input to the system end up in products, or in emissions to air, water or land. The quantities of these emissions must be estimated. The contents of rubbish bags, waste skips etc. have also to be examined and quantified.

It has been estimated (7) that over 50% of these wastes may be reclaimed, recycled or reused. Paper, cardboard and putrescibles have significant calorific values and may be combusted to produce heat and power, or may be converted to solid, liquid or gaseous fuels. When dumped to landfill, the combustible gas methane is produced over time, as well as the obnoxious gas hydrogen sulphide. Metals may be reclaimed and reused, glass containers may be reused, glass and minerals may be reclaimed, and textiles and plastics may be recycled or combusted under controlled conditions.

Table 7.2 Estimated total annual UK waste arisings

Sector	Annual arisings (million tonnes)	Date of estimate	Percentage of total arisings
Mining and quarrying	108	1989/1990	27
Agriculture	80	1991	20
Dredged spoils	43	1991	11
Sewage sludge	36	1991	9
Demolition/construction	32	1990	8
Household waste	20	–	5
Industrial waste	19	1990	4
Commercial waste	15	–	4
Other	50	–	12
Total	402		100

Table 7.3 Typical refuse analysis

Constituent	Percentage by weight	kg/household yr (4)
Putrescibles	30	120
Paper/cardboard	27	200
Dust	16	65
Glass	11	60
Metals	7	50
Textiles	3	20
Plastics	3	40
Minerals	3	15

Demand side options for cost-effective waste minimisation

Waste disposal options

The various options for waste disposal (5) are:

- elimination
- minimisation
- recycling
- reuse
- chemical treatments
- biological treatments
- composting, etc. of organic waste and sewage sludge
- incineration
- pyrolysis (heating to produce combustible gases)

- landfill
- disposal at sea
- disposal to foul sewer

BEST OPTION
Reuse
Eliminate
Reduce
Recycle
Treat
Dispose
WORST OPTION

Reuse

Bottles, containers and many types of packaging and objects may be reused, as are milk bottles and, in the quite recent past, as was the case for beer and soft drink bottles. No materials processing energy is needed between use and reuse, only washing and transportation.

Eliminate or reduce

Unnecessary waste may be eliminated by considering whether the product which is subject to disposal was really necessary in the first place, e.g. secondary packaging, unsolicited mail, free newspapers, circulars, etc.

Recycling

Direct recycling is the use of recycled materials instead of using virgin materials of the same type. Examples of this are the recycling of paper, glass, metals and plastic, resulting in less rapid depletion of raw material reserves.

Indirect recycling refers to the utilisation of a material for something other than its original purpose. Examples include the recycling of plastics into coarser plastics and the combustion of waste to release and use energy.

Whilst there are difficulties in separating the components of aluminium from mixed domestic waste, schemes such as the *'CROP' programme* in Milton Keynes, allowing the consumer to separate paper and card, plastics, metals and glass at source, are proving to be very popular.

Paper and card are constructed from cellulose, stemming mainly from softwoods, such as pine or spruce. Large quantities of energy and water are used in the manufacturing process and much pollution is produced. The total world consumption of paper and cardboard is of the order of 225 million tonnes/annum. Of this, approximately 30% comes from recycled pulp (8). It has been estimated that nearly 90% is capable of being retrieved for recycling. Paper married in a product to plastics, glues and metals cannot be recycled. During recycling, waste paper is formed into pulp, de-inked and then washed to remove the smallest fibres (*fines*) which weaken the paper produced. Paper cannot be recycled indefinitely (about four times at maximum) because, at each successive refining, or fibrillation process, where the fibres are teased from the pulp to promote bonding, the fibres are cut and shortened. Virgin pulp with long strong fibres must therefore always be mixed with recycled pulp, reducing the possibility that up to 90% of all paper may be recycled. Nevertheless, recycling reduces the amounts of raw materials, energy and water needed in the manufacturing process and the concomitant pollution of air, water and land.

Glass recycling is fairly straightforward and bottle banks are now commonplace. Glass collected is broken into cullet which can be used as a replacement for the raw minerals (70% sand (SiO_2), 16% limestone ($CaCO_3$) and 14% sodium carbonate (soda ash, Na_2CO_3)) in the manufacture of glass, with no deterioration of glass quality, as long as different colours and types of glass are segregated. There is therefore, in theory, no limit to the number of times that glass can be recycled, although present methods of manufacture allow only 50% cullet.

Glass recycling results in the need for:

- fewer raw materials
- fewer imports
- fewer quarries and their associated environmental impact
- less landfill space for the dumping of waste glass products
- less energy required for converting the raw constituents to an amorphous form (cullet melts at a lower temperature than sand)
- less energy-related environmental pollution
- less water for cleaning and other processing purposes

Metals that can be recycled include iron, tin and aluminium, particularly from drink cans, of which over 10 billion are used every year in the UK, contributing the major proportion of metal to the waste stream. It is worth reflecting that the use of cans for drinks is a relatively recent phenomenon—20 years ago there were no drink cans and, even today, their use is banned or discouraged in certain countries (e.g. Denmark).

Recycling *aluminium* saves 95% of the energy required to manufacture an aluminium can. Most methods for separating out aluminium waste from mixed refuse involve passing pulverised waste on a conveyer belt through alternating magnetic fields, which produce eddy fields in the metal. These react with the external magnetic field and produce a lifting force on the aluminium, removing it from the conveyer belt. The separation of *iron* involves passing the waste adjacent to a strong electromagnet.

Plastic bottles cannot be refilled and reused as it is not possible to sterilise them. They therefore have to be shredded and reprocessed for recycling. Plastics are not as easy to recycle as paper and glass since there are a wide variety of types, derived from oil and natural gas. The two main categories of plastic are:

- *thermoplastics*, which are the most commonly used in Europe (80%).These may be remoulded, and hence reused, if heated.
- *thermosetting* plastics which set once cooled and cannot then be reheated and remoulded.

Once used and combusted, plastics have similar calorific values to oil, but give off noxious and dangerous gases (e.g. dioxins and furans) if burnt under uncontrolled conditions. Plastics may, however, be treated safely by pyrolysis.

The current draft of an EC Directive requires 90% recovery of packaging and 60% recovery of material value. EC emission legislation is likely to insist that plastics be burnt in new incinerators which meet tight emission requirements.

Energy from waste

If it is not possible to reuse or directly recycle a material, then organic waste may be used as a source of energy. Some 30 million tonnes of such waste is discarded in the UK each year. This has a calorific value of about 12 MJ/kg or 3.33 kWh/kg (9). At a modest 2p per kWh, 10 000 million kWh of energy, worth £2 billion, could be produced each year (4% of the nation's fuel bill).

Bedfordshire, UK

The population of Bedfordshire, UK, was 541 200 (in 1991) and it has a land area of 123 000 hectares. Its commercial activities are dominated by agriculture and food processing. In addition to fruit and vegetables grown in the county, a great deal is also imported for processing to produce canned goods and juices, etc. In 1989/1990, the waste deposited to landfills contained 32 400 tonnes of paper, card, wood, etc., 765 800 tonnes of putrescible household waste and 38 900 tonnes of putrescible liquid waste.

When deposited to landfills, these wastes decay, producing the potentially explosive greenhouse gas methane and the obnoxious gas hydrogen sulphide. The nitrate-laden leachate from these wastes runs off into water courses to promote the growth of oxygen-hungry blue-green algae. This causes the eutrophication, or accelerated ageing, of streams, rivers and lakes by oxygen depletion, killing fish and other water-borne organisms.

This organic waste has a calorific value of around 12 MJ/kg. So the total of 837 100 tonnes/annum contains 10 000 million MJ, 2800 million kWh, which, at 2p/kWh, could produce £56 million worth of heat and power. This represents over 5000 kWh, worth £100 each to every man, woman and child in Bedfordshire, or £400 to the average home.

Compost derived from organic garden waste sells for £4 for 60 litres. The compost has a density of approximately 0.45 kg/litre and so is worth £150/tonne. The total annual amount of Bedfordshire's organic waste would be worth £133 million if composted. If crops were grown, using the nitrate-rich compost, fruit and vegetables to a value exceeding £500 million could be produced.

If fresh enough, this waste can be used directly for animal feed. It can also be neutralised in acidity, dried and pelleted for animal feed. Such feed is sold for £80/tonne. The total value of the solid putrescible waste as animal feed is thus £61 million. The bulk of the organic waste is, however, dumped to landfill.

Organic waste processing

Organic waste may thus be used as animal feed or to produce compost, aerobically, anaerobically or via vermocomposting. It may also be dried and combusted to produce heat for drying, heating greenhouses, cooking, sterilising or for district heating purposes. It can be used in a steam turbine installation to produce power and electricity. The electricity can be used to run lighting, process plant and equipment, such as refrigerating and freezing systems, or fed to the national grid.

It may be digested anaerobically or gasified by pyrolysis to produce methane gas or fermented to produce the liquid fuel, ethanol. These fuels may be used for sewage treatment, water purification, crop production in greenhouses (possibly using hydroponic bioculture) or fields, or other manufacturing processes.

Thus the waste organic material can be the sole input to a self-contained total energy system which produces animal feed, compost, chemicals, solid fuel, liquid fuel, gaseous fuel, electricity, water, crops and products.

Calorific values of organic wastes (9)

Nearly all organic waste fuels are cellulosic in character, and the heating value is a function of carbon and hydrogen contents. On a moisture-free basis, the heating values can be estimated at 8000 BTU/lb; more resinous materials at about 9000 BTU/lb.

Table 7.4 lists typical calorific values for various waste items.

Incineration of municipal solid waste (MSW):

- saves landfill space by reducing the volume of waste
- reduces transportation costs
- reduces costs for disposal
- reduces fossil fuel consumption if the energy released is utilised to produce electricity and/or hot water or steam for direct heating

Table 7.5 lists further calorific values for various waste products.

Fluidised bed combustion involves the passing of air through a bed of sand containing shredded waste. The whole becomes *fluidised*, enabling high heat transfer

Table 7.4 Typical calorific values for various waste items

By-product fuel	Calorific value MJ/kg (dry)	% moisture as received	Latent heat MJ/kg	'Wet' calorific value MJ/kg wet matter
Coffee grounds	23.26	65	1.63	6.51[1]
Cattle manure	17.2	50–75	1.25–1.88	1.35–2.42
Municipal refuse	22.1	43	1.15	11.45
Rice straw or hulls	13.9	7	0.18	12.75
Wheat straw	19.77	4	0.10	18.88
Corncobs	21.63	10	0.25	19.22

Notes

(1) 1 kg of wet material contains 35% of dry material, 0.35 kg having a calorific value of 23.26×0.35 = 8.14 MJ/kg, together with 0.65%, 0.65 kg of water which must be boiled off using 0.65×2.5 = 1.63 MJ, resulting in an overall calorific value of 8.14 – 1.63 = 6.51 MJ/kg of wet material.

Table 7.5 Calorific values for various waste products

Waste	Calorific value (MJ/kg)
Mixed plastic*	37
Polyethylene (PE)*	43
Polyvinyl chloride (PVC)*	22
Polystyrene*	38
Textiles*	16
Paper	15
Organic matter	7 (depending upon water content)
Municipal solid waste (MSW)	11 (depending upon water content)
Oil	44
Coal	28
Wood	16 (depending upon water content)
Heating oil	44

* beware of toxic and carcinogenic fumes.

rates to be achieved as the waste is combusted. Using this process better combustion is achieved and less pollution is produced. Limestone can be added to prevent sulphurous oxides entering the exhaust gas stream.

Refuse-derived fuels (RDF)

Solid fuel can be produced by separating the combustibles from the mixed waste and shredding these to form a floc. This floc can be pelletised into a safe and odourless storable and transportable solid fuel, having a calorific value of 16 MJ/kg.

Gaseous fuel can be produced via the anaerobic decomposition process that takes place in landfills over time, but also in more controlled conditions. Compacted waste contains little oxygen for aerobic decomposition, and so breakdown of the organic waste by anaerobic bacteria results in the production of the combustible gas methane (typically 50% methane and 50% carbon dioxide and so having a calorific value (22 MJ/kg) of about half that of natural gas). Each tonne of waste produces about 100 times its own volume of these gases, but may be converted to much lower volume storable and transportable liquefied natural gas substitutes. Gaseous fuel can also be produced via *pyrolysis*, which is the physical and chemical decomposition of waste brought about by the high temperature heating of waste in a limited supply of oxygen (i.e. in vacuum conditions). The process produces both gaseous and liquid emissions which can be used as fuels.

Liquid fuel can be produced by fermentation followed by distillation. Fermentation is the decomposition of organic substances by micro-organisms and/or enzymes (3). The process gives out heat and emits carbon dioxide as the sugars ($C_6H_{12}O_6$) convert to ethyl alcohol, or ethanol (C_2H_5OH), as follows:

$$C_6H_{12}O_6 = 2C_2H_5OH + 2CO_2$$

A kilogramme of sugar can be converted into half a kilogramme of alcohol in this way. The process is normally carried out under water. Since ethanol boils at 78°C, it must be separated from the water by distillation (evaporation of the alcohol at a temperature between 78.4°C and 100°C followed by condensation).

Thus the process converts the waste organic matter into a safe, odourless, storable and transportable high density energy form, having a calorific value of 30 MJ/kg.

Composting is the biological aerobic decomposition of organic wastes. Micro-organisms break down organic wastes into water, carbon dioxide and compost, which may be used to enrich soils. The decomposition is brought about by bacteria, fungi and actinomycetes, which, with suitable aeration, can raise the temperature of the compost to 70°C or more. At this temperature, all pests, seeds and pathogenic bacteria are destroyed, leaving a safe, odourless usable compost. If insufficient aeration is provided, the onset of anaerobic decomposition will produce biogas (methane, carbon dioxide and hydrogen sulphide) as happens in landfill.

Disposal

All materials that cannot be in any way reused or recycled should be treated and disposed of in a way that is least damaging to the environment. Incineration should be considered to reduce the volume of the waste (up to 90% is possible). This option should, however, be weighed against the additional air pollution produced and the reduction of propensive methane emitted and leachate leaking from future landfills.

The 39 steps to optimise waste management investments at a site

1. Identify site and scope of audit.
2. Obtain bills for purchased items for a recent representative year.
3. Obtain bills for waste disposal for a recent representative year.
4. Ascertain basis of charging for waste disposal.
5. Track the waste disposal route—landfill?
6. Investigate alternatives.
7. Estimate quantities of waste.
8. Estimate number of personnel on site.
9. Calculate waste quantity per person.
10. Compare with similar sites.
11. Examine and classify solid wastes:
    - combustible
    - compostible: biodegradable
    - paper and cardboard
    - glass
    - metals: tins and cans
    - textiles
    - plastics
    - ash
    - water content
    - disposable products
12. Examine and classify liquid wastes:
    - paints
    - solvents
    - oils
    - acids
    - alkalis
    - liquid fuels
13. Examine and classify gaseous and vaporous wastes:
    - CFCs
    - solvents
14. Record hazardous wastes:
    - chemical and common names
    - chemical and physical characteristics
    - physical and health hazards
    - possible routes of entry for humans/animals (inhalation, ingestion, skin or eye contact)
    - exposure limits and degree of carcinogenicity
    - precautions for handling and using
    - emergency first-aid required
    - supplier information

15. Pay attention to storage areas.
16. Investigate standards, regulations, Acts and legislation applicable.
17. Identify possible fines and penalties for illegal dumping.
18. Identify and cost needs for clean-up.
19. Identify options for reducing waste quantities:
    - is the product really necessary?
    - can packaging be reduced?
    - efficiencies of materials processing
20. Identify options for waste reclamation.
21. Identify options for waste recycling:
    - plastics
    - paper and cardboard
    - glass
    - metals
    - ash (e.g. as soil conditioner, fillers for road building)
    - biodegradables
    - pig bins
    - kerbside collection
22. Identify problems of collection and sorting waste.
23. Identify options for waste reuse: bottles, containers, clothes, etc.
24. Identify options for waste incineration: paper briquette production?
25. Examine uses for heat.
26. Examine possibilities for combined heat and power.
27. Examine possibilities for refuse-derived fuel production.
28. Examine possibilities for composting:
    - anaerobic
    - aerobic
    - vermocomposting
29. Examine possibilities for gasification via pyrolysis.
30. Examine possibilities for other gasification techniques.
31. Examine possibilities for hydrolysis/fermentation.
32. Analyses of options: straight rate of return.
33. Produce ranked investment portfolios.
34. Identify option conflicts.
35. Rerank investment portfolios.
36. Develop waste management strategy.
37. Establish five-year investment plan.
38. Set up monitoring and targeting and project management systems.
39. Project manage.

Rules for the efficient conservation of materials

The purposes for which expenditures of materials are required should be critically examined. As much useful purpose fulfilment should be extracted from a degrading materials chain, as is compatible with economic and other considerations.

The quality, not the quantity, of materials is the subject of conservation. Each materials-consuming operation should be examined critically and systematically in isolation and in relation to all other events occurring within the system boundary.

The manner and extent of all materials use should be challenged, including the appropriateness of the process method and the natures of the plant involved.

The following list of questions has been constructed from the perspective of the situation where an independent waste auditor is examining a client's site.

Questions for the client interview

Waste management procedures

- Who is responsible for waste management?
- Position in the organisation? Reporting to?
- Full or part time?
- Qualifications and experience?
- What is done?
- Has a waste audit been constructed?
- What has been achieved?

Financial practices

- Who controls the capital spending budget?
- Who controls the recurrent spending budget?
- Upon what financial criteria should cost-effectiveness calculations be performed?
- What is the period available to complete this exercise?
- Is there a list of waste-reducing investments under review, ranked in order of priority, with detailed costing and cost–benefit calculations?
- If not, why not?

Comments on waste quantities

- Is the amount of waste about right, too high, too low?
- What are the areas of high waste rejection?
- How much do you pay to dispose of waste?

Monitoring and recording practices

- How are the consumptions of materials reviewed?
- By whom?

- When was the last review?
- Is the waste analysed?
- What effluents and waste are generated on site?
  - solids
  - liquids
  - emissions to air
- What is the proportion of wastes to raw materials bought in?
  - by weight
  - by volume
  - by cost
- Where do the wastes go?
- What are the costs for waste disposal?
  - solids
  - liquid effluents
  - other
- Is there a materials consumption forecast/budget?
- Have standards been set? (i.e. for a given task or product or building)
- Is consumption compared with previous periods, other locations, other companies or other industries?
- What are the monitoring and targeting procedures?
- Should a materials and waste management policy be implemented?
- What should this policy be?

Personnel waste awareness

- Are details of materials consumption and waste rejection made known to employees?
- Are employees made aware of the need to reduce waste?
- What steps have been made to promote waste awareness via education and training, posters, etc.?

Current waste reduction measures

- What steps have been taken to reduce rates of waste rejection?
- What steps have been taken to reuse, reclaim or recycle waste materials?

Comments on inefficiencies

- Are there obvious incidents of unnecessary waste of materials?
- What are the monitoring procedures?
- What are the control arrangements?
- Are there leaks of liquid materials?

Major items

- What are the major waste-producing items of plant and equipment?

Waste storage systems

- What are the maintenance procedures?

Checklists for waste managers

Sources

- Materials should be used only when and where required
- Space or time delays inevitably incur losses
- Stocks should be maintained at minimum levels plus emergency reserves
- Attention should be paid to the delivery, storage and handling systems
- Comprehensive and accurate monitoring and metering of all materials inputs, throughputs and outputs should be accomplished
- A continuous log should be maintained
- Procedures should be standardised
- Qualities should be checked
- Information should be easily accessible, comprehensible and disaggregable
- Attempts should be made to account for all inputs in terms of outputs
- Storage areas should be made secure against loss or theft

Operations

- The grades of materials used should be matched to the purposes for which they are required
- The choice of a material to suit a particular application should not be made arbitrarily: different qualities are suitable for specific applications
- All losses and leaks of materials should be prevented
- Use recent deliveries of materials first
- Efficiency checks should be carried out frequently using standardised procedures
- Plant should be selected on sensible extreme conditions
- Materials cannot readily be conserved unless accurate and comprehensive measurements in consistent units are first obtained for all activities within the system boundary
- Greater overall efficiency can always be achieved at the cost of additional complexity
- Greater energy efficiency always requires an expenditure of materials and vice versa

- Side benefits and diseconomies—incidental benefits or penalties arising from each consuming activity should be identified and carefully evaluated
- Only the most efficient component branches in materials utilisation chains should be adopted
- Overall efficiencies are always lower than that of the most inefficient link in the chain
- Product designs should maximise lifespan, promote easy maintenance and repair, require little additional material inputs during active life, and should facilitate reuse, recycling, easy disposal and natural degradation and recycling

Pollution

- Waste and pollution should be closely monitored and minimised
- Waste in all forms not only squanders human effort, energy, time and materials, but also damages the external environment and disrupts ecological harmony
- Methods of waste collection, sorting and reclamation should be developed
- Improved recycling techniques are needed

The reduction of waste is especially desirable where materials have intense availability contents or where their historical availability costs are high. Metals, glass, plastics, paper and refractories are examples of such materials.

Design improvements which prolong the lifespan or promote the reuse or easy recycling of these energy-intensive materials are highly desirable.

Improvements

- Systems should be modelled and evaluated accurately so that the cost-effectivenesses of conservation options can be compared realistically
- Careful assessments of real savings should be made, including maintenance costs
- Full audits in common units should be carried out before and after improvements
- Evaluations should be obtained with respect to quantities of materials, financial costs of materials, and energy costs of materials
- Representative periods should be adopted for these analyses
- A continuous monitoring and targeting procedure should be implemented

The selection of new plant and processes for materials-conserving measures should be made, not on least capital cost criterion alone, but upon the basis of least total cost over the lifetime of the system.

Random factors should be eliminated—the system should be isolated from its external environment. All leaks and losses of materials should be prevented. Attempts should always be made to reduce demand before increasing materials supplies.

It should be ensured that modifications have no hidden diseconomic effects and that they comply with safety, fire and statutory regulations and codes of practice.

Reclamation

- The waste rejected from a high grade process might be collected and stored to be employed in another process
- Attempts should be made to recycle materials or incinerate waste
- Attempts to reduce or reuse waste should be made before any attempts at recycling or recovery
- Waste materials should be reused wherever economically possible, ensuring that practical grade-, time- and space-matched uses have been found for the reclaimed amounts
- The value of the savings must clearly exceed the cost of recovery
- Before attempting to recover 'waste', ensure that a matched need exists for it
- Consider the problems of collection, separation, storage and reprocessing
- Evaluate diseconomic effects (i.e. infestation, odours)

Controls

That which is not measured cannot be controlled.

Materials cannot readily be conserved unless accurate and comprehensive measurements in consistent units are first obtained for all activities within the system boundary.

Management

Great care should be taken to ensure the cleanliness, correct operation and planned systematic maintenance of all storage, release, distribution, utilisation and rejection systems.

Education

All personnel should be made fully aware of the waste implications of their activities and decisions.

All associated personnel should be availed of information, demonstrations of achievements and reports of failures of 'improvement' activities.

Laws of energy and materials flows

- Energy or matter can be neither created nor destroyed.
- Energy or matter are always conserved, although transductions may occur.

- Exergy degrades via equilibrium processes.
- All energy and materials tend to degrade to entropic disorder by being dispersed over a greater volume. Human beings' consumerism accelerates this process.
- As a result of utilisation, energy and materials are eventually downgraded to an equilibrium state having zero availability corresponding to the environmental datum.
- The environmental potential should be regarded as the reference datum.
- The availability of either energy or materials requires a source or sink at a potential different from that of the environment.
- The efficiency of an energy or materials conversion is always less than 100%.
- Decreasing entropy is a futile process in the long term but may be of temporary use in the short term.
- Energy must be lost in reducing entropy (work production).
- The supply of work can reduce entropy (heat pumping).
- Activities, if completely described, all lead to energy or materials rejection to the environmental datum, i.e. the total entropy of the considered control volume through which the energy or materials flow rises.
- Energy reclamation involves the expenditure of materials.
- Materials reclamation involves the expenditure of energy.
- Reclamation of energy or materials reduces pollution.

Reject side options

The general principles of BATNEEC are that solid wastes should be rendered harmless, packaged, clearly labelled and passed over to a registered waste disposal contractor.

Dust and ash should be bagged. Heavy metals may require special handling. All waste should be neutralised and stabilised as far as possible prior to disposal.

Some solid waste matter may be saleable (e.g. slag, sulphur, gypsum and waste catalyst). Instrumentation and compliance monitoring requirements include measurements of soil conditions and properties, local flora, prescribed substances and quantities and analyses of solid wastes.

7.7 IMPROVEMENT PROJECT

Having assessed, evaluated and ranked the various retrofit materials and waste-saving options in order of cost-effectiveness, the resulting improvement projects can be interleaved into the *energy and water* **investment portfolio**, the **project plan** and the **project management** and **monitoring and targeting programmes**.

REFERENCES

(1) P.W. O'Callaghan, *Energy Management*, McGraw-Hill, Oxford, 1993.
(2) J. Elkington and J. Hailes, *The Green Consumer Guide*, Victor Gollancz, London, 1988.
(3) A. Porteous, *Dictionary of Environmental Science and Technology*, Open University Press, Milton Keynes, 1991.
(4) P.H. Collin, *Dictionary of Ecology and the Environment*, Peter Collin Publishing, Teddington, 1988.
(5) *Croners Environmental Management*, Croner Publications, Surrey, 1991.
(6) Department of the Environment, *Digest of Environmental Protection and Water Statistics*, HMSO, No. 15, 1992.
(7) A. Brown, *The UK Environment*, Department of the Environment, Government Statistical Office, London, 1993.
(8) M.O. Shone, Waste Reclamation and Waste Auditing, MSc Thesis, Cranfield Institute of Technology, 1992.
(9) E.A. Avallone and T. Baumeister, *Marks Standard Handbook for Mechanical Engineers*, Ninth Edition, McGraw-Hill, New York, 1987.
(10) WARMER Factsheet, *Fuel From Waste*, The World Resource Foundation, Tonbridge, UK, January, 1991.

8

Integrated case study: a corporate site

This chapter contains a fully worked integrated environmental management case study of a corporate site. The 39 steps of environmental management, energy management, water management and waste management have been worked through systematically. Thirteen good investment options to save energy, water and materials have been identified. Eight of these options involve zero capital cost and the remaining five options have financial payback periods of less than two years. By adopting a no-risk investment plan, where perceived and realised savings are reinvested in further cost-effective options, annual financial savings of £320 363 are possible for no net capital investment. In addition to this, the project plan indicates that a sum of £442 109 of accumulated savings plus interest will have accrued at the end of the five-year project period. This case study demonstrates the importance of the integrated environmental management approach, as it allows more investment options to be adopted earlier in the project, resulting in maximising financial savings during the duration of the project. The summarised results of similar studies, involving 34 companies in total, show that similar savings are possible for a range of industrial and commercial sectors. It is therefore possible for a company simultaneously to comply with environmental legislation, adopt high environmental standards, save resources, save money and protect the environment.

Considering energy, water/effluent and waste separately, results in final bills of 95% (energy), 73% (water/effluent) and 26% (waste) of the original bills, resulting in annual savings at the end of the project period of £70 900 and accumulated savings of £285 690.

When all three sectors are considered together, the final bills become 67% (energy), 73% (water/effluent) and 26% (waste) of the original bills, resulting in annual savings at the end of the project period of £320 363 and accumulated savings of £442 109.

8.1 ENERGY AND ENVIRONMENTAL AUDIT, GREEN PRODUCTS LTD.

Executive summary

Main recommendations

- Short seminars should be set up for management employees to highlight current laws and directives pointing to future legislation and the environmental impacts of the company's activities, and to describe the operation of energy and environmental monitoring and targeting systems.
- Workshops should be held on environmental issues for supervisory staff and shopfloor employees to foster responsibilities for departmental environments.
- An environmental manager should be appointed.
- A monitoring and targeting system should be set up.
- An electricity maximum demand monitoring and control system should be instituted, as well as maximum demand scheduling by coordinating the operation of large machines.
- The use of electricity for weather-dependent heating purposes should be investigated.
- Procedures should be introduced whereby unnecessary lights are switched off.
- Lights and luminaires should be replaced with low energy devices.
- An *input/output* analysis should be performed over the boiler system.
- Compressed air pressure should be reduced to 90 psi.
- Furnace operations should be rescheduled to off-peak times.
- The ventilating air system should be examined.
- An energy management system should be installed.
- Water distribution systems should be visually inspected.
- Leaks from water mains should be repaired.
- Dripping taps should be fitted with new washers.
- Infra-red sensors should be installed at urinals.
- The labelling of partially used metal stocks should be improved.
- Attention should be paid to the storage and disposal of special wastes.
- Paper waste should be minimised.
- Cost-effective means of organic waste processing should be sought.
- The more efficient use of metals should be sought.

Thirteen good investment options have been identified to save energy, water and materials with less than a two-year payback on investment. Eight of these options involve zero capital cost.

By adopting a no-risk investment plan, where perceived and realised savings are reinvested in further cost-effective options, annual financial savings of £320 363 are possible *for no net capital investment*. In addition to this, the project plan indicates that a sum of £442 109 of accumulated savings plus interest will have accrued at the end of the five-year project period.

8.2 MAIN REPORT

39 steps of integrated environmental management

IEM Step 1: Identify education and training needs

In order to promote environmental awareness throughout the company, Green Products Ltd. needs to address energy conservation, water conservation, materials conservation, waste prevention and environmental legislation in its employee education and training programmes.

All employees should be made aware of the environmental impacts of their activities and their environmental responsibilities. Customers, suppliers, creditors and investors should also be made aware of the environmental issues involved in their activities.

IEM Step 2: Set up training programmes

Short seminars should be set up for management employees to highlight current laws and future directives and the environmental impacts of the company's activities, and to describe the operation of energy and environmental monitoring and targeting systems.

Workshops should be convened on environmental issues for supervisory staff and shopfloor employees to foster responsibilities for departmental environments.

Brief environmental seminars should be held showing the company's commitment to the environment, customers, suppliers and stakeholders.

IEM Step 3: Make a statement of company commitment

'Green Products Ltd. regards the protection of the environment as a high priority in its affairs and is committed to protecting the environment by supplying the necessary resources and personnel to fulfil this aim. The company will comply with environmental legislation and conform to high environmental standards, and at the same time, will strive to minimise the use of energy, water, and other natural resources, whilst reusing and recycling materials wherever economically possible.'

IEM Step 4: List environmental issues

The environmental issues of concern to Green Products Ltd. are:

- high energy usage

Table 8.1 SWOT analysis for Green Products Ltd.

Strengths	**Weaknesses**
Environmental commitment	Lack of monitoring and targeting
Funding committed	Energy-intensive processes
Good reputation	Lack of education in environmental matters
Profitable company	Lack of environmental training
Well-established	Reactive approach
Opportunities	**Threats**
Great potential for energy savings	Environmental legislation
Potential for materials savings	Rising utility bills
Greater competitiveness	Local pressures

- emissions to air: oxides of nitrogen, oxides of sulphur, particulates, volatile organic compounds
- emissions to land/water: leachates of oils and chemicals
- thermal pollution: from steam discharges
- noise emissions
- excess packaging

IEM Step 5: Conduct an initial review

Table 8.1 contains the SWOT analysis for Green Products Ltd.

IEM Step 6: Set company policy

Points of policy:

- to be more efficient
- to conduct audits and monitoring
- to budget for environmental improvements
- to reduce deleterious environmental impact
- to comply with legislation and possibly do better than the law demands
- to comply with standards
- to adopt an integrated approach to environmental management
- it should cover suppliers, the company and its customers
- it should be supported from the highest level in the company
- it should be understood by all employees
- it should be made publicly available

IEM Step 7: Set company objectives

(1) To reduce emissions to air, land and water to realistically achievable levels using cost-effective demand side options and reject side options in accordance with BATNEEC and other guidelines, within a given time frame.
(2) To establish a monitoring and targeting system so that compliance with the above objective can be measured.
(3) To produce a list of prioritised actions, with dates of implementation.

IEM Step 8: Set company targets

Targets and dates for these objectives will be set up as a result of this integrated environmental management study.

IEM Step 9: Construct a company policy statement

'Green Products Ltd. is committed to reducing its environmental impact through minimising harmful emissions to air, land and water, by employing appropriate energy, water and materials conservation methods, and by reducing visual and audible pollution. By making the most efficient use of resources, and continually reviewing, assessing and refining environmental performance, Green Products Ltd. will lead by example in their actions towards environmental improvements.'

IEM Step 10: Make an environmental statement

'Green Products Ltd. manufactures products designed to protect the environment. Because of its product range and position in the marketplace, it is particularly conscious of its need to be "squeaky green" in its activities.'

In 1994, Green Products Ltd. spent:

- £763 966 on gas and electricity
- £28 883 on potable water supply
- £28 763 on drainage of surface water
- £7562 on sewage and effluent charges
- £200 on the disposal of oil/water emulsions
- £16 200 on solid waste disposal
- £750 on bin rental

making £846 324 in all.

Relevant *environmental issues* are:

- emissions to air
- emissions to water
- emissions to land

Green Products Ltd. is striving to reduce these emissions via promoting:

- the efficient use of energy
- the efficient use of water
- the efficient use of materials

and by introducing end-of-pipe clean-up facilities as are reasonable given the types and magnitudes of the emissions and the economic implications.

A monitoring and targeting programme is under way to measure and assess the environmental improvements made.

IEM Step 11: Structure the environmental organisation

The environmental management team, under the direction of the site environmental manager, comprises three full-time personnel:

Person	**Responsibility**
Ian Burns	Energy, emissions to air, water and effluents
Irma Dumper	Materials, packaging, waste and transport
E.S. Milud	Environmental law, health and safety

IEM Step 12: Construct a register of regulations

The following documents have been identified and are housed at the company's environmental management centre's display area.

Regulations

- 1991 EC Eco-audit regulation
- 1991 Environmental Protection (Authorisation of Processes) Regulations
- 1991 Environmental Protection (Applications, Appeals and Registers) Regulations
- 1991 Environmental Protection (Duty of Care) Regulations
- 1991 Environmental Protection (Prescribed Processes and Substances) Regulations
- 1989 Control of Pollution (Registers) Regulations
- 1989 Control of Pollution (Consents for Discharges) Regulations
- 1989 Control of Industrial Pollution (Registration of Works) Regulations
- 1989 Trade Effluents (Prescribed Processes and Substances) Regulations
- 1989 Air Quality Standards Regulations
- 1988 Collection and Disposal of Waste Regulations
- 1984 Control of Pollution (Consents for Discharges) Regulations
- 1983 Health and Safety (Emissions into the Atmosphere) Regulations
- 1980 Control of Pollution (Special Waste) Regulations

Acts
- 1991 Water Resources Act
- 1990 Environmental Protection Act
- 1990 Planning (Hazardous Substances) Act
- 1989 Control of Smoke Pollution Act
- 1989 Water Act
- 1989 Control of Pollution (Amendment) Act
- 1988 Environment and Safety Information Act
- 1978 Refuse Disposal (Amenity) Act
- 1974 Health and Safety at Work Act
- 1974 Control of Pollution Act
- 1973 The Water Act
- 1968 Clean Air Act
- 1956 Clean Air Act

IEM Step 13: Assess compliance with legislation

This had not previously been done for Green Products Ltd., as the company was not aware of new legislation or how it would affect the company.

IEM Step 14: Perform an environmental compliance audit

As a matter of urgency, the environmental management team left the comfort of the environmental management centre to undertake a detailed examination of the site, identifying the following:

- Discarded drums of orimulsion—special waste
- Discarded containers of trichloroethane—carcinogenic solvent
- Discarded containers of trichloroethylene—carcinogenic solvent
- Discarded containers of used sump oil
- Discarded electrical transformers—special waste containing highly toxic PCB residues

An authorised waste disposal contractor was immediately summoned to dispose of these offending items, which were handed over with the exchange of appropriate dispatch notes, labels and documentation.

The drums were eventually sold to a new orimulsion-fired power station nearby.

IEM Step 15: Carry out scoping study

Full energy, water and waste audits will be undertaken for the site.

IEM Step 16: Determine the objectives of the work to be done

(1) *To conserve resources by using less energy, water and input materials, and, as a result, save money.*
(2) *To ensure compliance with environmental legislation and so avoid fines and other penalties.*
(3) *To adopt high environmental standards, to be seen to be 'green' and so enhance competitive advantage.*

IEM Step 17: Set the targets for the work to be done

To identify and prioritise cost-effective investment options for environmental improvement.

IEM Step 18: Construct an environmental management programme

The auditors will track the flows of energy, water and materials via *input*, *throughput* and *output* analyses.

IEM Step 19: List the action plan

For energy, electricity, water and materials:

(1) Analyse bills.
(2) Examine points of release.
(3) What are the commodities used for?
(4) Where do they end up?
(5) How can less be used?
(6) How can emissions be cleaned up?
(7) Generate prioritised list of improvement options.
(8) Construct investment plan and project schedule.
(9) Set up monitoring and targeting system.
(10) Undertake project.

IEM Step 20: Construct an activity schedule

Week Activity
(1) Analyse bills.
(2) Examine points of release.
(3) What are the commodities used for?
(4) Where do they end up?
(5) How can less be used?
(6) How can emissions be cleaned up?

(7) Generate prioritised list of improvement options.
(8) Construct investment plan and project schedule.
(9) Set up monitoring and targeting system.
(10) Undertake project.

IEM Step 21: Identify information requirements

The energy, water and waste audits will require analyses of:

- fuel bills
- electricity bills
- water bills
- bills for waste
- bills for effluent disposal
- bills for materials bought in

Also needed are:

- site plans
- plans of site services
- buildings details
- UA values and glazing ratios for the buildings
- number of employees
- working schedules
- production figures
- types of heating and heaters used

IEM Step 22: Set up an environmental management system

This comprises the documentation developed in Steps 1 to 21 and the register of environmental effects, which follows.

IEM Step 23: Document the register of environmental effects

(1) *Energy consumption and emissions to air*. This will be tackled via energy audits and recommendations for energy conservation, as well as investigating methods of exhaust gas clean-up.
(2) *Oil contamination in water discharge*. Sources of oil will be investigated and preventive measures implemented.
(3) *Refrigerant contamination in water discharge*. Sources of refrigerant contamination will be investigated and preventive measures implemented.
(4) *Leachate from externally stored materials*. This has been dealt with via the removal of the offending materials. Remaining materials are now stored

under cover on a solid concrete base. All such stored materials have been contained in appropriate packaging and are now clearly labelled.

(5) *Materials consumption and waste to land.* All uses for materials throughout the site are under investigation via a materials audit. Opportunities for reuse, reclaimation and recycling are being sought. Unavoidable waste is packaged, clearly labelled and consigned to an authorised waste disposal contractor.

(6) *Use of solvents.* The uses for solvents are being investigated. Methods are being sought to reduce the amounts used. Escape channels are being identified. Whilst as a result of containment there are no direct leaks to land or water, the main problem is evaporation, leading to emissions to air. Methods of containment are being investigated, as well as the use of alternative safe solvents.

(7) *Use of land: visual impact.* There are substantial opportunities for landscaping and the addition of flora to the site. Such improvements will be added to the investment plan, to be resourced from financial savings resulting from energy, water and materials conservation measures.

(8) *Noise.* Sources of unacceptable noise are being investigated and suitable noise attenuation/suppression systems are being installed.

(9) *Odours.* Sources and emissions of odours are being investigated and means of suppression are being sought.

(10) *Products.* All products are now subject to 'cradle-to-grave' environmental life cycle analyses. This includes quantification of environmental impacts in the following chain:

- mining materials
- extraction of materials
- materials packaging and its disposal
- transportation of materials
- materials processing
- storage of materials
- manufacture of products
- services (e.g. building services, overhead functions)
- product packaging and its disposal
- distribution of the product
- retailing the product
- use of the product
- disposal of the product
- reuse/recycleability of the product

IEM Step 24: Compile an environmental management manual

This will be constructed from this report.

IEM Step 25: Perform an environmental review

The aim of the environmental review is to construct a snapshot of the environmental performance of the company.

It involves *input/throughput/output* accounts of all energy, water and waste streams flowing through the site.

This step incorporates:

- the 39 steps of energy management
- the 39 steps of water management
- the 39 steps of materials and waste management

as follows.

39 steps of energy management

EM Step 1: Identify the site and boiler/building combinations

The site contains various production processes, such as metals jointing, machining and heat treatment in ovens.

Space heating is supplied by free-standing gas-fired radiant heaters and gas-fired hot air blowers. There is also some electrical heating. A small gas-fired boiler provides low pressure hot water (LPHW) for office radiators.

EM Step 2: Obtain fuel and electricity bills for a recent representative year

Table 8.2 lists the energy forms delivered over the annual period considered.

EM Step 3: Obtain degree-days for the location

Table 8.2 includes the degree-days for the region in which the site is located.

Table 8.3 lists the prices of gas, electricity, electrical maximum demand and supply capacity, and Table 8.4 lists the resulting costs of energy delivered.

The mean cost of electricity, taking into account maximum demand and supply capacity charges, is 9.6p/kWh.

The consumption of natural gas accounted for 39% of the energy bill and electricity cost the remaining 61%, with 10% of the total being paid as maximum demand supply capacity charges.

EM Step 4: Identify thermal or other changes that have occurred over or since the year under examination

No significant changes occurred.

Table 8.2 Energy delivered (units)

Month	Gas (1E6 kWh)	Electricity (1E6 kWh)	MD (kVA)	SC (kVA)	'Excess' SC	Degree-days	Days/ month	MMT (°C)
Jan	2.5	0.5	1400	1600	200	308	31	5.6
Feb	2.1	0.5	1300	1600	300	310	28	4.4
Mar	2.0	0.5	1500	1600	100	283	31	6.4
Apr	1.2	0.4	1150	1700	550	193	30	9.6
May	0.4	0.35	1000	1700	700	141	31	11.6
Jun	0.35	0.3	750	1700	950	60	30	14.7
Jul	0.15	0.25	1100	1700	600	50	31	13.5
Aug	0.15	0.25	1000	1700	700	71	31	12.8
Sep	0.3	0.3	1200	1700	500	115	30	11.7
Oct	1.5	0.5	1235	1700	465	242	31	4.3
Nov	2.1	0.5	1300	1700	400	335	30	7.7
Dec	2.2	0.5	1350	1700	350	324	31	5.1
	14.95 total	4.85 total	1500 max.	1700 max.	5815 total	2432 total	365 total	8.95 mean

Notes
1E6 ≡ $\times 10^6$
MD—electrical maximum demand
SC—electrical supply capacity
MMT—mean monthly temperature

Table 8.3 Prices

Gas	2p/kWh	
Electricity	8p/kWh	
Month	**MD (p/kVA)**	**SC (p/kVA)**
Jan	1200	60
Feb	1200	60
Mar	400	60
Apr	90	60
May	90	60
Jun	60	60
Jul	60	60
Aug	60	60
Sep	60	60
Oct	60	60
Nov	400	60
Dec	1200	60

Table 8.4 Costs of energy delivered (£)

Month	Gas	Electricity	MD	SC
Jan	50 000	40 000	16 800	960
Feb	42 000	40 000	15 600	960
Mar	40 000	40 000	6 000	960
Apr	24 000	32 000	1 035	1 020
May	8 000	28 000	900	1 020
Jun	7 000	24 000	450	1 020
Jul	3 000	20 000	660	1 020
Aug	3 000	20 000	600	1 020
Sep	6 000	24 000	720	1 020
Oct	30 000	40 000	741	1 020
Nov	42 000	40 000	5 200	1 020
Dec	44 000	40 000	16 200	1 020
Total	299 000	388 000	64 906	12 060
%	39	51	8.5	1.5

Grand total cost of energy: £763 966.

EM Step 5: Examine electrical supply capacity, maximum demand variations and all tariffs

Table 8.2 shows that, after reaching a peak maximum demand of 1500 kVA in March, the electrical supply capacity was increased to 1700 kVA, costing an extra £720 per annum. Yet 1500 kVA was never exceeded during the representative year. The 'excess' supply capacity, 5815 kVA, cost the company £3489 during the year. If the electrical load can be controlled via peak lopping and the like, the maximum demand and the supply capacity might be reduced. Any electrical energy savings will also result in lower monthly peak maximum demand charges and allow a reduction in supply capacity. The company should, in any case, apply immediately to reduce the supply capacity to 1500 kVA, saving £1440 in a full year. Large electrical energy users should be identified so that these might be switched off as the peak demand is approached (e.g. large machines, air compressors, ventilation fans, banks of lights or electrical water heaters). Daily variations in maximum demand should be examined to identify when the peaks occur.

Recommendation 1: Introduce a maximum demand control system.

There is also a substantial variation in maximum demand over the year, indicating that high-premium electricity is being used for weather-dependent heating purposes. If this heating requirement were eliminated and supplied from gas-heated appliances, the electrical maximum demand and the supply capacity might be further reduced.

The implications of this possibility are examined in Table 8.5.

Table 8.5 Energy delivered (common units)

Month	Gas (1E6 kWh)	Electricity (1E6 kWh)	Mean (kW)	MD (kVA)	SC (kVA)	'Excess' (SC)
Jan	2.5	0.5	672	1400	1600	200
Feb	2.1	0.5	744	1300	1600	300
Mar	2	0.5	672	1500	1600	100
Apr	1.2	0.4	556	1150	1700	550
May	0.4	0.35	470	1000	1700	700
Jun	0.35	0.3	417	750	1700	950
Jul	0.15	0.25	336	1100	1700	600
Aug	0.15	0.25	336	1000	1700	700
Sep	0.3	0.3	417	1200	1700	500
Oct	1.5	0.5	672	1235	1700	465
Nov	2.1	0.5	694	1300	1700	400
Dec	2.2	0.5	672	1350	1700	350
	14.95 total	4.85 total	744 peak	1500 peak	1700 peak	5815 total

Mean kW in Table 8.5 is calculated by dividing the electrical kWh consumed each month by the hours in the month. This figure never exceeds 800 kVA and shows a significant variation over the year. The lowest monthly amounts, 0.25E6 kWh, were consumed in July and August. If this is the electrical baseload for the site, it becomes 3E6kWh in a full year, costing £240 000. If the remaining 1.85E6 kWh, costing £148 000, were due to electrical heating at 8p/kWh, £84 360 per annum might be saved by switching to gas-fired systems at 2p/kWh.

The baseload, 0.25E6 kWh, corresponds to a mean kW of 342 kW. Thus, if the electrical load were reduced to the baseload, the supply capacity could be reduced to 400 kVA and the alternative figures emerge as indicated in Table 8.6.

This results in the costs given in Table 8.7. This reduces the bills from £763 966 to £595 277, resulting in an annual saving of £168 689.

Recommendation 2: Investigate the use of electricity for weather-dependent heating purposes.

EM Step 6: Feed data to a spreadsheet

EM Step 7: Convert data to common units

The data were supplied in common units.

EM Step 8: Convert degree-days to mean monthly temperatures during operating hours

This was not necessary as mean monthly temperatures were supplied, as listed in Table 8.2.

Table 8.6 Energy delivered (common units)

Month	Gas (1E6 kWh)	Electricity (1E6 kWh)	MD (kVA)	SC (kVA)	'Excess' SC
Jan	2.75	0.25	336	400	64
Feb	2.35	0.25	336	400	64
Mar	2.25	0.25	336	400	64
Apr	1.35	0.25	336	400	64
May	0.5	0.25	336	400	64
Jun	0.4	0.25	336	400	64
Jul	0.15	0.25	336	400	64
Aug	0.15	0.25	336	400	64
Sep	0.35	0.25	336	400	64
Oct	1.75	0.25	336	400	64
Nov	2.35	0.25	336	400	64
Dec	2.45	0.25	336	400	64
	16.8 total	3.0 total	336 peak	400 peak	768 total

Table 8.7 Costs of energy delivered (£)

Month	Gas	Electricity	MD	SC
Jan	55 000	20 000	4032	240
Feb	47 000	20 000	4032	240
Mar	45 000	20 000	1344	240
Apr	27 000	20 000	302.4	240
May	10 000	20 000	302.4	240
Jun	8 000	20 000	201.6	240
Jul	3 000	20 000	201.6	240
Aug	3 000	20 000	201.6	240
Sep	7 000	20 000	201.6	240
Oct	35 000	20 000	201.6	240
Nov	47 000	20 000	1344	240
Dec	49 000	20 000	4032	240
Total	336 000	240 000	16 397	2880
Percentage	56.5	40	3	0.05

Grand total costs of energy: £595 277.

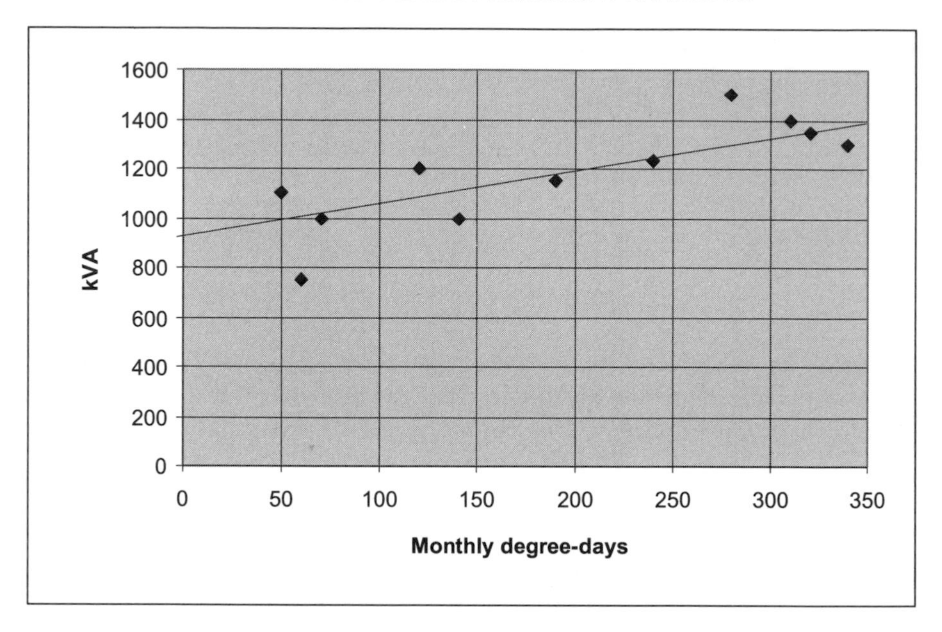

Figure 8.1 Electricity maximum demand versus degree-days

EM Step 9: Plot kVA versus degree-days

This is shown in Figure 8.1.

EM Step 10: Plot kVA versus mean monthly temperatures during operating hours

This step is not necessary here as the plant operates for 24 hours a day.

EM Step 11: Correlate with linear regression and quantify scatter

The monthly maximum demand kVA contained in Table 8.5 were first correlated with the degree-days listed in Table 8.2, using linear regression (see Table 8.8).

It was found that the data conformed with:

$$\text{kVA} = 880 + 1.53 \times \text{D}$$

EM Step 12: Identify electricity use for heating

Superimposed about the regression line in Figure 8.1 are the data for each month. It may be seen that, whilst there is scatter in the data (the standard error of the estimate being 122 kVA), there is a definite trend for the maximum demand to follow outside climatic conditions. The 550 kVA is demanded to supply this variable load.

Table 8.8 Maximum demand versus degree-days

Degree-days (DD)	kVA	Regression output	
308	1400	Constant	880.1559
310	1300	Std err of Y est	122.5676
283	1500	R squared (Adj)	0.6466149
193	1150		0.6787408
141	1000	No. of observations	12
60	750	Degrees of freedom	10
50	1100	Coefficient(s)	1.530892
71	1000	Std err of coef.	0.3330584
115	1200		
242	1235		
335	1300		
324	1350		

Table 8.9 Maximum demand scenario

Month	MD (kVA)	SC (kVA)	MD cost (£)	SC cost (£)
Jan	850	900	10 200	510
Feb	850	900	10 200	510
Mar	850	900	3400	510
Apr	850	900	765	510
May	850	900	765	510
Jun	850	900	510	510
Jul	850	900	510	510
Aug	850	900	510	510
Sep	850	900	510	510
Oct	850	900	510	510
Nov	850	900	3400	510
Dec	850	900	10 200	510
Total	850 peak	900 constant	41 480	6120

Assuming that the baseload is 850 kVA and eliminating the climate-dependent demand results in the scenario shown in Table 8.9.

This results in a reduced bill for maximum demand and supply capacity from £76 966 to £47 600, which is probably a more realistic possibility.

EM Step 13: Plot kWh of electricity versus degree-days

This is shown in Figure 8.2.

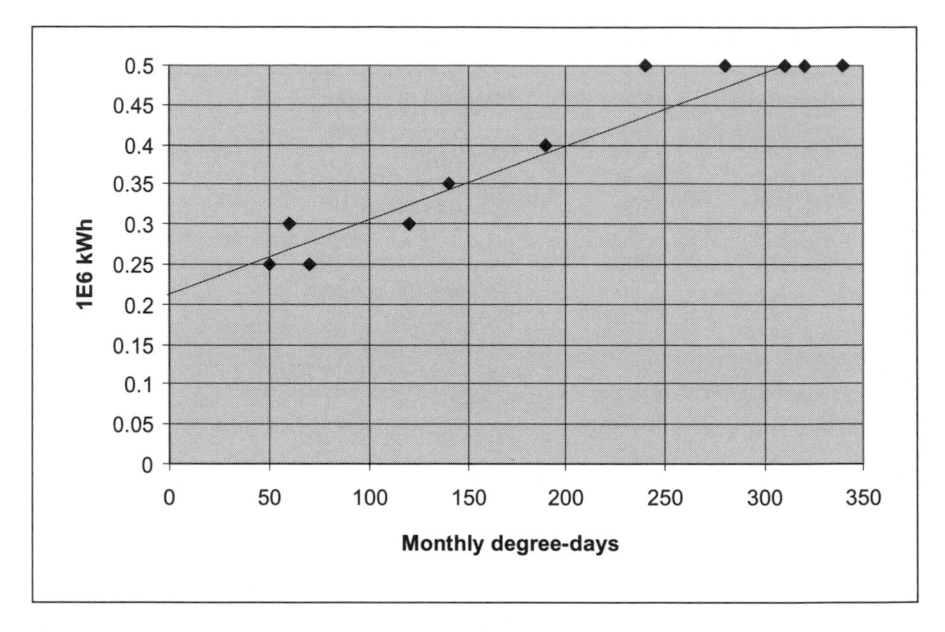

Figure 8.2 Electricity kWh versus degree-days

Table 8.10 Regression analysis: electricity consumption versus degree-days

Degree-days per month	kWh electricity per month (1E6 kWh)	Regression output	
308	0.5	Constant	0.2135646
310	0.5	Std err of Y est	0.0273990
283	0.5	R squared (adj)	0.9351271
193	0.4	(raw)	0.9410246
141	0.35	No. of observations	12
60	0.3	Degrees of freedom	10
50	0.25	Coefficient(s)	0.0009405
71	0.25	Std err of coef.	0.0000745
115	0.3		
242	0.5		
335	0.5		
324	0.5		

Table 8.11 'Heating' electricity

Degree-days per month	'Heating' kWh electricity per month (1E6 kWh)
308	0.27
310	0.27
283	0.27
193	0.17
141	0.12
60	0.07
50	0.02
71	0.02
115	0.07
242	0.27
335	0.27
324	0.27
Total	2.09

It was found that the data conformed with

$$\text{Electricity kWh/month} = (0.214 + 0.00094 \times \text{DD/month}) \times 1\text{E}6$$

Figure 8.2 indicates that the scatter of the data about this regression line is small. The baseload for electrical use is 0.23E6 and the dependence upon outside climatic conditions is clear.

EM Step 14: Quantify electricity use for heating

Subtracting the baseload from the data results in the data given in Table 8.11

Figure 8.2 clearly shows that electricity is being used for heating purposes. This costs £167 200 per annum. If this electrical heating were eliminated and gas substituted, £125 400 per annum might be saved. The baseload, 2 760 000 kWh/annum, results in a constant mean monthly maximum demand of 315 kVA, in line with the figure derived earlier. Proper maximum demand control might then reduce the monthly maximum demands appreciably and so allow the supply capacity to be reduced.

Recommendation 3: Investigate electrical load patterns and seek means to accomplish maximum demand control.

EM Step 15: Plot kWh of electricity versus mean monthly temperature over the operating hours

EM Step 16: Correlate with linear regression and quantify scatter

As the plant is on 24-hour operation, these steps are not necessary.

EM Step 17: Check electricity use for heating

A detailed survey of electricity utilisation should be undertaken, in line with Recommendation 2.

EM Step 18: Obtain annual mean furnace/boiler efficiency to discount fuel supplied to obtain heating energy

The instantaneous boiler efficiency was measured to be 86%. Whilst there were no records relating gas input to hot water output, it was noted that the boiler was cycling (frequently turning on and off). A modulated control system might therefore improve the seasonal performance of the boiler. Boiler operation should be monitored over a representative period in order to assess the overall annual efficiency and to estimate the financial savings that may possibly result from boiler control.

Recommendation 4: Perform input/output analysis over the boiler system.

EM Step 19: Plot kWh of heating energy versus degree-days

In order to ascertain the total heat supplied for space-heating purposes, the electricity identified as being used for heating must be added to the heat supplied from the boiler, as accomplished in Table 8.12.

Again, the total heat supplied for space heating is correlated with degree-days in Table 8.13.

The data conform to the expression:

$$\text{Total heat supplied} = (-0.359 + 0.0079184 \times \text{degree-days}) \times 1E6$$

Figure 8.3 indicates a very good correlation between heating energy and degree-days, the scatter in the data being small.

Notice that the regression line crosses the x-axis at around 50 degree-days. Thus the base temperature for the site is lower than the base temperature upon which the degree-days are based (15.5°C).

The base temperature for the site is defined as that value of outside air temperature which internal sundry gains (arising from non-heating internal electricity use, human metabolic activity, solar gains and other internal activity) lift to the level of the internal air temperature (1).

Table 8.12 Total heat supplied for space heating

Degree-Days per month	'Heating' kWh electricity per month 1E6 kWh	Gas supplied to boiler per month 1E6 kWh	Heat supplied from boiler per month 1E6 kWh	Total heat supplied 1E6 kWh
308	0.27	2.5	2.15	2.42
310	0.27	2.1	1.806	2.076
283	0.27	2	1.72	1.99
193	0.17	1.2	1.032	1.202
141	0.12	0.4	0.344	0.464
60	0.07	0.35	0.301	0.371
50	0.02	0.15	0.129	0.149
71	0.02	0.15	0.129	0.149
115	0.07	0.3	0.258	0.328
242	0.27	1.5	1.29	1.56
335	0.27	2.1	1.806	2.076
324	0.27	2.2	1.892	2.162
2432 total	2.09 total	14.95 total	12.857 total	14.947 total

Table 8.13 Correlation of total heat supplied for space heating with degree-days

Degree-Days per month	Total heat supplied (1E6 kWh)	Regression output	
308	2.42	Constant	−0.358970
310	2.08	Std err of Y est	0.1985175
283	1.99	R squared (adj)	0.9512136
193	1.2	(raw)	0.9556487
141	0.46	No. of observations	12
60	0.37	Degrees of freedom	10
50	0.15	Coefficient(s)	0.0079184
71	.15	Std err of coef.	0.0005394
115	.33		
242	1.56		
335	2.08		
324	2.16		

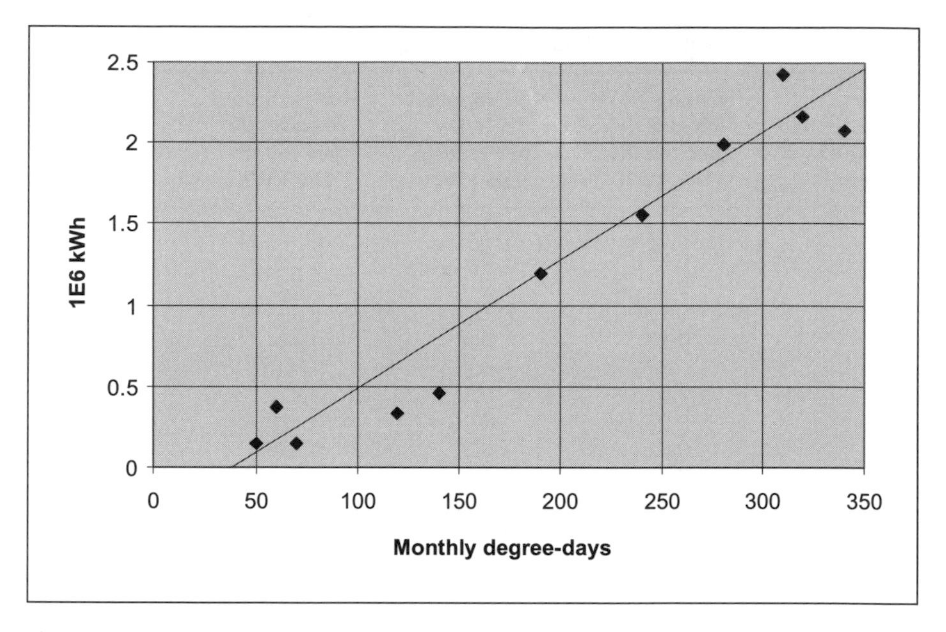

Figure 8.3 Heating energy versus degree-days

EM Step 20: Plot kWh of heating energy versus monthly mean air temperatures

It is this relationship, the *energy signature*, which reveals the value of the *base temperature* for the site.

EM Step 21: Correlate with linear regression and quantify scatter

Table 8.14 performs the correlation of hcating energy versus monthly mean air temperatures.

The correlation suggests that the data may be represented by:

$$\text{Heat supplied} = (3.28 - 0.227 \times \text{MMT}) \times 1\text{E}6 \text{ (kWh)}$$

EM Step 22: Obtain annual mean inside air temperature

The site aimed to maintain the inside air temperature at 17°C.

EM Step 23: Obtain base temperature

From Figure 8.4, the base temperature for the site (i.e. the outside air temperature pertaining when no heating or cooling is required) is 14.45°C.

Table 8.14 Correlation of monthly heating energy versus monthly mean air temperatures

MMT (°C)	Heat supplied (1E6 kWh)	Regression output	
5.6	2.42	Constant	3.275397
4.4	2.08	Std err of Y est	0.2562100
6.4	1.99	R squared (adj)	0.9187368
9.6	1.2	(raw)	0.9261244
11.6	0.46	No. of observations	12
14.7	0.37	Degrees of freedom	10
13.5	0.15	Coefficient(s)	−0.226767
12.8	0.15	Std err of coef.	0.0202533
11.7	0.33		
7.7	1.56		
4.3	2.08		
5.1	2.16		
8.95	14.95		
Mean	Total		

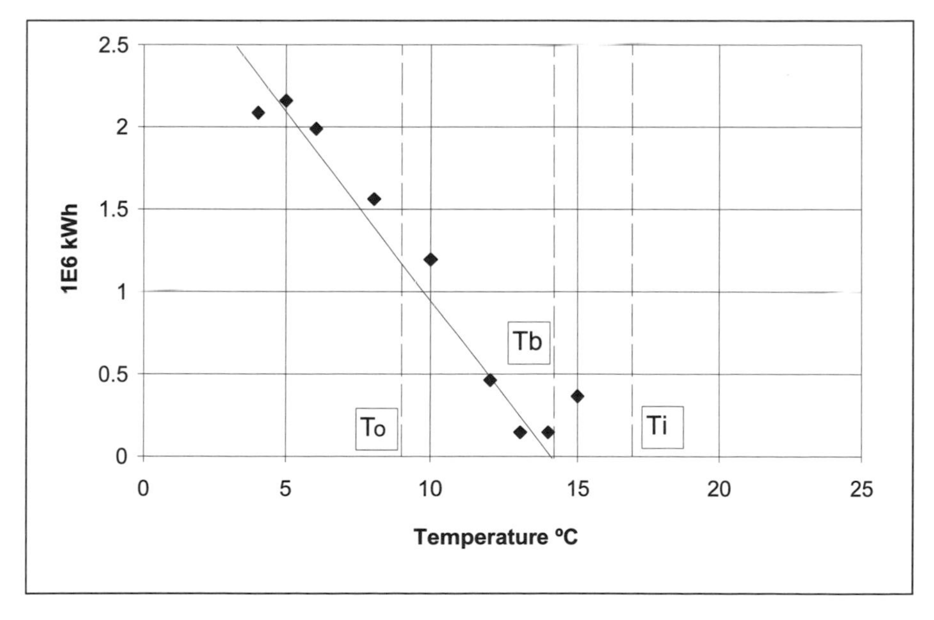

Figure 8.4 The energy signature: monthly heating energy versus mean monthly outside air temperatures

EM Step 24: Quantify sundry gains and direct rejects of energy

From reference (1):

Annual internal sundry gains = $(UA + nV/3) \times (T_i - T_b) \times$ (hours per year)/ 1000
Annual heating energy = $(UA + nV/3) \times (T_b - T_o) \times$ (hours per year)/1000
where
U ($Wm^{-2}K^{-1}$) is the effective overall U-value for the buildings at the site, A (m^2) is the exposed surface area of all the buildings at the site, n (hr^{-1}) is the effective overall number of fresh air ventilation airchanges that the buildings experience and V (m^3) is the total enclosed volume of the site buildings.
Dividing and cancelling terms,
Sundry gains/heating energy = $(T_i - T_b)/(T_b - T_o)$
= 0.458
Now, the total heat delivered annually = 14.95 E6 kWh
and so the annual sundry gains are 6.85 E6 kWh.

EM Step 25: Obtain equation relating the overall U-value to the number of airchanges, n

A site survey produced the following:

Total exposed area of the buildings = 35 000 m^2
Total enclosed volume = 36 000 m^3
From the heating energy equation,
viz
Annual heating energy = $(UA + nV/3) \times (T_b - T_o) \times$ (hours per year)/1000
$14.95E6 = (35000\ U + 12000\ n) \times 5.6 \times 8760/1000$
$304753.8 = 35000\ U + 12000\ n$
or
$n = 25.39 - 2.92\ U$

EM Step 26: External site survey

An extensive site survey (1) resulted in the calculation of a site overall U-value of 3.9 $Wm^{-2}\ K^{-1}$.

EM Step 27: Deduce the number of airchanges, n

Substituting this U-value in the above equation yields

$$n = 13.0 \text{ airchanges per hour}$$

The *CIBSE Guide* (2) recommends only three airchanges per hour for this type of site, and so there are considerable losses via uncontrolled ventilation.

Great financial gains might be made, therefore, by paying attention to reducing n, via weather-stripping and the like.

Each airchange costs annually £11 754 in gas, or £47 015 in electricity (plus maximum demand and supply capacity charges), and so the 10 extra airchanges per hour cost the company £117 540 each year!

Recommendation 5: Investigate ventilating air system.

EM Step 28: Construct output analysis

Outputs =
Fuel energy losses during combustion, conversion and distribution
+ electrical energy directly rejected during utilisation
+ fabric transmission losses
+ ventilation heat losses

$0.14 \times$ gas energy input
+ non-heating electricity
$+ 35\,000U \times 5.8 \times 8760/1000$
$+ 12000n \times 5.8 \times 8760/100$

= 2 093 000
+ 2 760 000
+ 6 696 144
+ 7 926 048

= 19 475 192 kWh

EM Step 29: Investigate other fuel-derived effluents and stack losses

The main production areas were fitted with extraction fans to extract fumes. These fans were causing the huge number of airchanges throughout the plant. By partitioning off certain high fume-producing areas, ducting outside air to these isolated areas and extracting air and fumes via the exhaust fans (through a balanced flue), the airchanges in the rest of the factory could be reduced.

Apart from the stack emissions from the boilerhouse, no other emissions to air, or effluents to land or water, were noted.

EM Step 30: Internal surveys if process plant and equipment significant

Whilst the production processes use small but significant amounts of energy, the lack of internal monitoring made it impossible to analyse these.

Table 8.15 The energy audit balance sheet for the site

	kWh	£
Inputs		
Gas	14 950 000	299 000
Electricity	4 850 000	388 000
SC/MD		76 966
Totals	19 800 000	763 966
Outputs		
Flue gas losses	2 093 000	41 860
Electrical devices	2 760 000	220 800
Fabric transmission losses	6 696 144	133 923
Ventilation losses	7 926 048	158 521
Totals	19 475 192	513 243.8
Residual	324 808	
(from rounding errors)	(1.64%)	

EM Step 31: Construct input, output, throughput analysis

All the ingredients necessary for the energy audit balance sheet have now been assembled.

EM Step 32: Construct energy audit balance sheet

Table 8.15 is the energy audit balance sheet for the site.

Notice that the *input* and *output* costs do not balance. This is because SC/MD costs do not result in output costs, and because electricity at 8p per kWh is being used to offset fabric transmission and ventilation losses, which should be served by gas at 2p per unit.

EM Step 33: Identify options for energy conservation and money saving

Summary of initial recommendations:

- Recommendation 1: Introduce a maximum demand control system.
- Recommendation 2: Investigate the use of electricity for weather-dependent heating purposes.
- Recommendation 3: Investigate electrical load patterns and seek means to accomplish maximum demand control.
- Recommendation 4: Perform input/output analysis over the boiler system.
- Recommendation 5: Investigate ventilating air system.

EM Steps 1 to 32 have already revealed that significant savings may be made as follows:

(1) *Elimination of electrical heating and the substitution of gas heaters.* Electrical heating (2.09E6 kWh/annum) costs the company £167 200 per annum. Substituting gas would cost £48 604 (taking the boiler inefficiency into account), saving £118 595 per annum.

Electrical heating also incurs an additional cost in electrical maximum demand and supply capacity charges. Each kW lopped off the maximum demand produces a reduction of 1kVA in maximum demand charges; 2.09E6 kWh/annum lopped off the electricity usage reduces maximum demand by 238kVA.

The existing maximum demand and supply capacity charges are as listed in Table 8.16.

If 238 kVA were lopped off the maximum demand and 400 kVA from the supply capacity, the scenario listed in Table 8.17 would result, saving £14 314 in a full year.

The total savings resulting from the adoption of this recommendation then become £132 909 per annum.

(2) *Weather-stripping the building, the introduction of vestibules and the like, partitioning off and separately venting fume-producing zones.* The number of fresh airchanges is excessive at 13 airchanges per hour. This should be reduced to 3 airchanges per hour, saving £117 540 (or £136 674 taking the boiler inefficiency into account).

(3) *The installation of maximum demand monitoring and control.* By perfect load levelling, the maximum demand could be reduced to 342kVA with the supply capacity set at 400 kVA, resulting in the 'best' scenario listed in Table 8.18.

This indicates that a total further saving of £43 082 might be possible with this option.

(4) *Paying attention to lighting.* Lighting levels were measured at a number of sampling points throughout the site. These varied from 170 to 740 lux and were often found to be excessive for the task illuminated. Luminaire substitution could result in annual savings of about £10 000.

Unoccupied areas were invariably found illuminated. Switching off unnecessary lights may save £10 000 per annum.

The total saving possible from paying attention to lighting is then £20 000.

(5) *Paying attention to compressed air services.* This is a most expensive energy commodity. A system pressure reduction from 100 psi to 90 psi would save £650 per annum. There was much evidence of compressed air leakage from pipes and fittings which should be given attention.

Note: Options (4) and (5) will also result in further electrical maximum demand and supply capacity charges.

Table 8.16 Existing maximum demand and supply capacity charges

Month	MD (kVA)	MD (p/kVA)	Cost (£)	SC (kVA)	SC (p/kVA)	Cost (£)	MD and SC (£)
Jan	1400	1200	16 800	1600	60	960	17 760
Feb	1300	1200	15 600	1600	60	960	16 560
Mar	1500	400	6000	1600	60	960	6960
Apr	1150	90	1035	1700	60	1020	2055
May	1000	90	900	1700	60	1020	1920
Jun	750	60	450	1700	60	1020	1470
Jul	1100	60	660	1700	60	1020	1680
Aug	1000	60	600	1700	60	1020	1620
Sep	1200	60	720	1700	60	1020	1740
Oct	1235	60	741	1700	60	1020	1761
Nov	1300	400	5200	1700	60	1020	6220
Dec	1350	1200	16 200	1700	60	1020	17 220
Total			48 706			12 060	76 966

Table 8.17 Adjusted maximum demand and supply capacity charges

Month	MD (kVA)	MD (p/kVA)	Cost (£)	SC (kVA)	SC (p/kVA)	Cost (£)	MD and SC (£)
Jan	1162	1200	13 944	1300	60	780	14 724
Feb	1062	1200	12 744	1300	60	780	13 524
Mar	1262	400	5048	1300	60	780	5828
Apr	912	90	820.8	1300	60	780	1600.8
May	762	90	685.8	1300	60	780	1465.8
Jun	512	60	307.2	1300	60	780	1087.2
Jul	862	60	517.2	1300	60	780	1297.2
Aug	762	60	457.2	1300	60	780	1237.2
Sep	962	60	577.2	1300	60	780	1357.2
Oct	997	60	598.2	1300	60	780	1378.2
Nov	1062	400	4248	1300	60	780	5028
Dec	1112	1200	13 344	1300	60	780	14 124
Total			39 948			9360	62 652

EM Step 34: Analysis of options—straight rate of return

EM Step 35: Produce a ranked investment portfolio

Table 8.19 presents the portfolio of *energy* investment options for Green Products Ltd., ranked in order of rate of return on invested capital.

EM Step 36: Identify option conflicts

Table 8.18 'Best' maximum demand and supply capacity charges

Month	MD (p/kVA)	MD (£)	Cost (kVA)	SC (kVA)	SC (p/kVA)	Cost (£)	MD and SC (£)
Jan	1200	4104	342	400	60	240	4344
Feb	1200	4104	342	400	60	240	4344
Mar	400	1368	342	400	60	240	1608
Apr	90	307.8	342	400	60	240	547.8
May	90	307.8	342	400	60	240	547.8
Jun	60	205.2	342	400	60	240	445.2
Jul	60	205.2	342	400	60	240	445.2
Aug	60	205.2	342	400	60	240	445.2
Sep	60	205.2	342	400	60	240	445.2
Oct	60	205.2	342	400	60	240	445.2
Nov	400	1368	342	400	60	240	1608
Dec	1200	4104	342	400	60	240	4344
Total		12 586				2880	195 670

Table 8.19 Energy investment portfolio

Option	Description	Capital cost (£)	Savings per year (£)	Straight rate of return (%)
A	Compressed air	0	650	infinity
B	Lighting switch-off	0	10 000	infinity
C	Lighting replacement	5000	10 000	200
D	MD control	20 000*	20 000	100
E	Electrical heating	180 000	132 909	74
F	Ventilation control	136 674	80 000	58

Note
* conservative estimate.

EM Step 37: Rerank the investment portfolio

There are no apparent option conflicts, although the audit should be repeated as each new set of recommendations is implemented.

EM Step 38: Establish five-year investment plan

The five-year investment plan for *energy* retrofit measures is listed in Table 8.20 and the summary of energy savings is given in Table 8.21.

EM Step 39: Set up monitoring and targeting and project management systems

To ensure success, actual and target spends on energy commodities and investments should be tracked monthly.

Table 8.20 Five-year investment plan

Year	Option	Capital cost (£)	Accumulated savings (£)
1	A and B	0	10 650
2	C	5000	15 650
3			26 306
4	D	20 000	56 950
5			97 600

Table 8.21 Energy summary (£)

Initial annual energy bill	763 966
Final energy bill	723 316 (95%)
Gross capital invested	25 000
Net capital invested	zero
Accumulated savings	97 600

39 steps of water management

WAT Step 1: Identify site and scope

Green Products Ltd.

WAT Step 2: Obtain drainage, effluent and sewerage bills for a recent representative year

- Surface water drainage costs: £28 763 in 1993.
- Trade effluent and measured sewage charges: £7 562 in 1993.

WAT Step 3: Obtain water bills for a recent representative year

- Cost of water: 50p per m^3.
- Number of shift personnel: 600.

The monthly water bills are listed in Table 8.22.

WAT Step 4: Examine water supply and effluent capacities, demand variations and bases of charging

The quantities and costs of water supplied are included in Table 8.22.

Table 8.22 Water bills

Month	Water supplied (m^3)	Cost of water supplied (£)	Days per month	Water per capita day	Degree-days per month	Water per degree-day (m^3/DD)
Jan	3280	1640	31	0.18	308	10.65
Feb	5340	2670	28	0.32	310	17.23
Mar	5260	2630	31	0.28	283	18.59
Apr	6620	3310	30	0.37	193	34.30
May	6020	3010	31	0.32	141	42.70
Jun	5412	2706	30	0.30	60	90.20
Jul	3708	1854	31	0.20	50	74.16
Aug	3836	1918	31	0.21	71	54.03
Sep	4162	2081	30	0.23	115	36.19
Oct	4124	2062	31	0.22	242	17.04
Nov	4150	2075	30	0.23	335	12.39
Dec	5854	2927	31	0.31	324	18.07
	57 766	28 883	365	0.26	2432	23.75
	Total	Total	Total	Mean	Total	Mean

WAT Step 5: Obtain degree-days data for the location

These are also included in Table 8.22.

It may be seen that the supply of potable water cost Green Products Ltd. £28 883 in the year considered. There is little dependence on degree-days and so the variations are dependent on the water use by production processes.

WAT Step 6: Obtain rainfall data for the location

As the company had no plans for rain water harvesting, this step was not necessary.

WAT Step 7: Identify changes in water usage that have occurred over or since the year of examination

No significant changes had occurred.

WAT Step 8: Trace and construct diagram of water handling and distribution systems

These were not available.

WAT Step 9: Trace and construct diagram of surface drainage, sewage and effluent handling, distribution and disposal systems

Works drainage plans were obtained.

WAT Step 10: Estimate number of personnel on site—calculate water/person

- Number of personnel: 600.
- Water per annum: 57 766 m^3.
- Water per annum per person = 96.28 m^3/annum/person = 0.26 m^3 per person per day.

WAT Step 11: Quantify monthly sewage quantities

Bills are charged according to rateable value.

WAT Step 12: Feed data to a spreadsheet

WAT Step 13: Plot water use versus months of the year

WAT Step 14: Obtain mean monthly outside air temperatures

WAT Step 15: Plot water use versus degree-days

WAT Step 16: Plot water use versus mean monthly outside air temperatures

WAT Step 17: Correlate with linear regression and quantify scatter

Figure 8.5 shows the variation of water costs over the months of the year. Because the unit cost for water was invariant over the period considered at 50p/m^3, it was not necessary to plot and correlate water quantities. The relationship shows no dependence on outside climate conditions.

WAT Step 18: Perform site survey

During the site survey of water usage, it was noted that water-cooled equipment and the use of water in urinals were by far the main consumers of water.

WAT Step 19: Measure water qualities throughout

Mains quality potable water is used throughout the site.

WAT Step 20: Quantify different quality water uses

The water-cooled test equipment was used sporadically throughout the year. Figure 8.5 shows that the water baseload for the site is probably 3600 m^3 per month, or 43 200 m^3 in the full year. The bulk of this baseload was used in the automatic

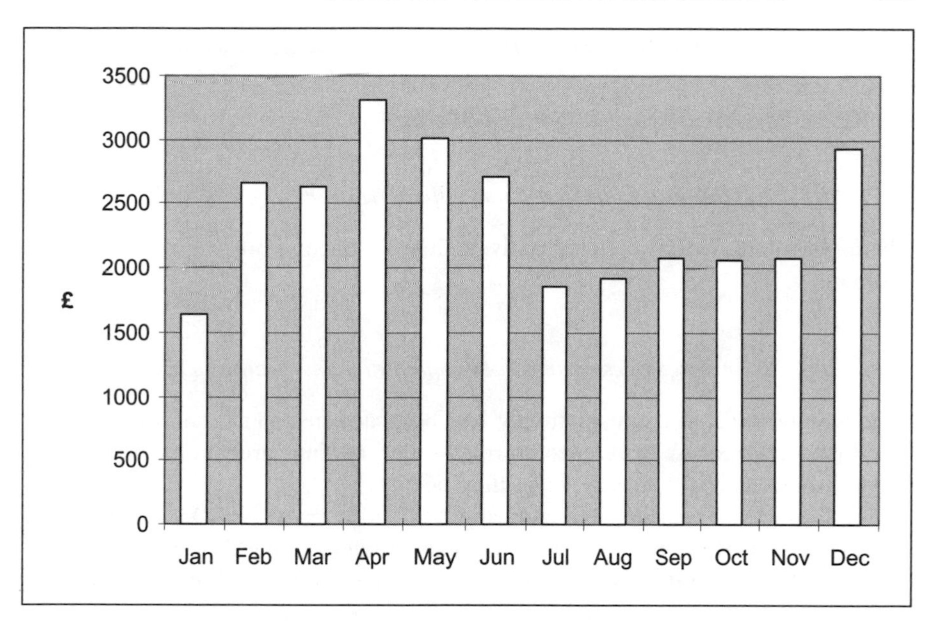

Figure 8.5 Variation of water charges over the months of the year

flushing of urinals. The remaining (57 766 – 43 200) 14 566 m^3 of water was used to cool the test equipment.

WAT Step 21: Construct output analysis—where does all the water end up?

From the previous step it is concluded that 75% of the water ended up as sewage from the urinals.

WAT Step 22: Quantify evaporated water

It was estimated that about half the cooling water was lost through evaporation and that a further 10% of total water might be lost via seepage.

WAT Step 23: Look for evaporation prevention possibilities

This was considered, but the complexity of the cooling systems on the equipment made the installation of evaporation prevention or capture systems not economically feasible.

WAT Step 24: Investigate and measure effluents, quantities and qualities

WAT Step 25: Investigate standards, regulations, acts and legislation applicable

WAT Step 26: Identify possible fines and penalties

WAT Step 27: Identify and cost needs for effluent clean-up

There were fortunately no significantly polluted effluents, other than the sewage outflows.

WAT Step 28: Investigate losses and leaks—quantities and qualities lost

Serious water leaks in the main supply had been detected some months after they had first occurred. A water monitoring and targeting programme should be established to highlight such leaks as they occur.

The flushing frequencies of urinals appeared to be excessive. One such system was found to be flushing every two minutes throughout the 24 hours. Infra-red detectors should be fitted to all urinal flush systems. Since the legal minimum for flushing cisterns is once every 20 minutes, the sensor systems must also incorporate over-ride timer controls.

Waste leaking from taps in consequence of having faulty washers was also identified as a source of waste. A three-monthly inspection and maintenance programme should be instituted for all water-using devices—taps, toilets, washing machines, etc.

WAT Step 29: Internal surveys if water-using process plant and equipment significant

The use of water for equipment cooling should be further investigated.

WAT Step 30: Construct throughput analysis

Table 8.23 presents the water throughput analysis.

WAT Step 31: Construct water audit balance sheet—quantities and qualities

Due to discrepancies between figures of water supplied and water consumed, this needs further detailed investigation.

WAT Step 32: Identify options for water conservation and money-saving options

The annual water bill is £28 883. Table 8.24 presents the water accounts.

In view of the high proportion of potable water used for the flushing of urinals

Table 8.23 Water throughput analysis

Inputs	Mains water
	Rain water
Throughputs	Processes
	WCs
	Washing
	Cooking
	Drinking
Outputs	Evaporation
	Leaks to sewer
	Seepage to land
	Burst mains
	Discharges

Table 8.24 Water accounts (m^3)

	Quantity	%
Input		
Water delivered	57 766	100
Uses		
Cooling equipment	14 566	25
Urinals	43 200	75
Other	negligible	
Outputs		
Sewage	43 200	75
Evaporation	7283	12.5
Seepage	5776	10.0
Other	1444	2.5

and costing £21 600 per annum, the management should consider reversing its decision not to investigate the use of rain water for flushing urinals.

Possible areas for savings are:

- to reduce demand
- to monitor consumption on a weekly basis to identify trends and leaks
- to introduce water cascading—it may be possible to use certain cleaner effluent discharges for equipment cooling; since this happens sporadically throughout the year, a storage (settling) tank may be required
- to reduce leaks via inspection and maintenance
- to install infra-red sensors to minimise flushings

WAT Step 33: Analyses of options—straight rate of return

Table 8.25 Water investment portfolio

Option	Description	Capital cost (£)	Savings per annum (£)	Straight rate of return (%)
A	Visual inspection of systems	negligible	1400	infinity
B	Washers on dripping taps	negligible	150	infinity
C	Infra-red sensors at urinals	1500	16 000	1066

WAT Step 34: Produce ranked investment portfolio

Table 8.25 presents the portfolio of *water* investment options for Green Products Ltd., ranked in order of rate of return on invested capital.

There are therefore opportunities to save £17 550 per annum for an investment of only £1500. In addition, opportunities for water cascading should be sought and rain water harvesting for flushing toilets should be considered.

WAT Step 35: Identify option conflicts

There are no option conflicts.

WAT Step 36: Rerank investment portfolio

This is not necessary.

WAT Step 37: Establish five-year investment plan

The five-year investment plan for *water* retrofit measures is listed in Table 8.26, and the summary of water savings is given in Table 8.27.

With the reduction in water use, a case might be made to the Water Disposal Authorities for a reduction in effluent disposal charges.

WAT Step 38: Set up water monitoring and targeting system

WAT Step 39: Project manage

39 steps of waste management

WAS Step 1: Identify site and scope

Green Products Ltd.

Table 8.26 Five-year investment plan

Year	Option	Capital cost (£)	Accumulated savings (£)
1	A, B and C	1500	16 050
2			33 600
3			51 150
4			68 700
5			86 250

Table 8.27 Water summary (£)

Initial annual water/effluent bill	65 208
Final water/effluent bill	47 658 (73%)
Gross capital invested	1500
Net capital invested	zero
Accumulated savings	86 260

WAS Step 2: Obtain bills for purchased items for a recent representative year

There were difficulties in obtaining bills for materials purchased, but stock items cost in the region of £1 million and materials cost the company of the order of £10 million in the year considered.

WAS Step 3: Obtain bills for waste disposal in the representative year

The removal of waste bins and oil/water effluent cost the company £17 150 during the year.

WAS Step 4: Ascertain basis of charging for waste disposal

- *Bins*—fixed rate £90 per bin collected plus £50/bin charge for hire of bins
- *Oil/water*—£200 charge per 1000 gallons

WAS Step 5: Track the waste disposal route

Bins to landfill. Oil/water unknown, but consignment notes exchanged.

WAS Step 6: Investigate alternatives

Alternative contractors are selected by tender annually. The company did not wish to consider alternative methods of waste disposal.

WAS Step 7: Estimate quantities of waste

Number of bins	15 (rental £750/annum)
Capacity/bin	6 m^3
Collections/year	12
Waste quantity per annum	936 m^3
Cost per bin	£90
Unit cost	£15 per m^3
Cost per 180 bin collections	£16 200
Oil/water effluent	3.4 m^3
Cost of effluent disposal	£200
Total cost	£17 150

No information concerning types of solid waste could be obtained from the company.

WAS Step 8: Estimate number of personnel on site

600 persons.

WAS Step 9: Calculate waste quantity per person

$$936/600 = 1.56 \text{ m}^3 \text{ per annum} = \text{approximately } 780 \text{ kg per annum}$$
$$= 2 \text{ kg per person-day}$$

This does not appear to be excessive. It was noted, however, that a wastepaper recycling scheme, conducted by a local charity, operated in the offices. This could discount the total waste significantly, but as the wastepaper was given over to the charity free of charge no records of quantities were kept.

WAS Step 10: Compare with similar sites

There were no other similar sites available for comparison.

WAS Step 11: Examine and classify solid wastes

Sample bins were examined and these were found to contain the following, all mixed together:

Domestic waste:
- food packaging
- waste food
- clothing

- about 50% paper documents
- shredded documents

Building materials:
- insulation
- bricks
- wood

The company may benefit from waste separation, i.e. separating the waste into separate bins for:

- organic waste
- combustible waste
- aggregate
- textiles
- other waste

The waste oil/water mixtures were stored in underground tanks and were suspected to contain solvents and refrigerants.

The use of metals in the manufacturing processes caused huge quantities of steel plate offcuts and angle iron to be discarded. This requires in-depth investigation.

Other waste identified included:

- *scrap metal*—collected by a scrap merchant
- *special wastes*—trichloroethylene, orimulsion (now handed over to a specialist waste disposal contractor)
- *glass*—a small amount
- *plastics*—a small amount in packaging materials
- *ash*—some piles of ash were evident in the waste storage area; care must be taken to ensure that this does not leach to the water table

WAS Step 12: Examine and classify liquid wastes

Apart from the orimulsion and trichloroethylene, liquid waste was generally evident in small quantities. It included:

- paints
- solvents
- oils/water
- acids mixed with alkalis to form pure brine solutions
- liquid fuels

WAS Step 13: Examine and classify gaseous and vaporous wastes

The main gaseous emissions to air were:

- stack losses
- solvents entering the extract ventilation systems—this should be investigated; the amount and type of solvents purchased should be obtained from the purchase orders

WAS Step 14: Record hazardous wastes

The processes operated do not produce significant quantities of hazardous wastes. The stored orimulsion and trichloroethylene have been dealt with.

The company had recently engaged specialist contractors to remove all remaining asbestos lagging on site and an unspecified number of discarded electrical transformers, previously left to lie in the waste storage area and presenting the risk of seepage of PCB (polychlorinated bi-phenols), had also been removed by specialist contractors.

WAS Step 15: Pay attention to storage areas

The waste storage area contained:

- skips for waste metals
- piles of ash and other particulates
- discarded building materials

In particular, the ash and grit should be disposed of to a licensed waste disposal contractor.

WAS Step 16: Investigate standards, regulations, acts and legislation applicable

The company considers that, having implemented a number of clean-up measures, it is operating in compliance with the law and is now considering application for BS 7750 and approval by the EU Eco-audit standard.

WAS Step 17: Identify possible fines and penalties for illegal dumping

There was no evidence of illegal dumping at the site.

WAS Step 18: Identify and cost needs for clean-up

All that remains now is the removal of the ash and grit. Quotations from contractors are being obtained.

WAS Step 19: Identify options for reducing waste quantities

The major cost savings are potentially to be made from the use of metals, costing by far the greatest proportion of the annual bill for materials. A series of seminars has been set up for employees dealing with the conservation of resources—energy, water and materials. Designers, production managers and external consultants meet regularly to seek the optimal use of metals and other materials for waste minimisation. The purchasing and stock control procedures are being revised and computerised so that stock quantities are minimised according to production requirements.

Product packaging had been reduced considerably and major crating and cartoning was returnable.

WAS Step 20: Identify options for waste reclamation

Consideration should be given to using some of the waste metal discarded. Project leaders have been asked to consider this.

WAS Step 21: Identify options for waste recycling

- *Paper.* The introduction of information technology has resulted in a huge increase in the amount of high quality computer paper, which is presently shredded and discarded to the waste disposal bins. Consideration should be given to the use of erasable printing papers and erasing printers. *Are the printouts really necessary when the information is much more rapidly accessed through the computer screens?* Discussions have been initiated with the information technology manager on the possibility of introducing a networked system with shopfloor access. These two options are in conflict.
- *Metals.* It is not possible to recycle metals on site. The greatest gains will come from optimising the use of raw materials. The information technology manager is seeking software which will enable templating and metal cutting to be optimised.

WAS Step 22: Identify problems of collection and sorting waste

Whilst no problems of waste collection exist, the separation of waste may present problems, viz. the separation of ferrous and the various non-ferrous metals. The scrap value is considerably increased for separated metals. At present, the scrap merchant does not pay for the metals removed. With a good separation system, an alternative scrap metals handler would pay for the metal waste. Alternatively, the metals could be sold back to the refineries.

Organic kitchen waste can easily be separated out, as can paper and cardboard, aggregate and combustible waste. A local use for the latter needs to be identified.

WAS Step 23: Identify options for waste reuse

There were no significant possibilities, although some bottles and containers might be reused. Most of the larger packaging items are already being reused.

WAS Step 24: Identify options for waste incineration

Although there are no possibilities for incineration on site, a local charity manufactures paper briquettes for distribution to old-age pensioners to supplement coal for fires. They should be approached and offered discarded waste paper and card.

WAS Step 25: Examine uses for heat

As the site is supplied with natural gas, all heating requirements can be provided relatively cheaply from the gas boilers. Any on-site combustion of waste would cause environmental problems.

WAS Step 26: Examine possibilities for CHP (combined heat and power)

A gas-fired CHP system might be considered when the gas boilers need to be replaced. The boilers have, however, at least a 16-year life remaining.

WAS Step 27: Examine possibilities for RDF (refuse-derived fuel) production

This is not considered to be within the remit of the company, except for the paper briquette production dealt with in WAS Step 24.

WAS Step 28: Examine possibilities for composting

A local environmental group in conjunction with the landfill company are considering setting up a composting facility to cope with organic waste. This will reduce nitrate leachate and methane emissions from future landfills. Green Products Ltd. is considering a partnership with these two parties for the manufacture of a range of organic waste-processing plants, including composters, fermenters to produce liquid ethanol and anaerobic digesters. In the meantime, some of the separated organic waste from the kitchens will be processed at the company's research centre in the prototype organic waste-processing equipment being developed there.

WAS Step 29: Examine possibilities for gasification via pyrolysis

WAS Step 30: Examine possibilities for other gasification techniques

The organic materials are possibly better dealt with via the techniques described in WAS Step 28.

WAS Step 31: Examine possibilities for hydrolysis/fermentation

This is being undertaken at the research centre.

WAS Step 32: Analyses of options: straight rate of return

The oil/water tanks are emptied once per year for a (minimum) cost of £200 for 1000 gallons. During the year considered, the tanks contained only 750 gallons of the mixture. If collection were delayed until the tanks were nearly full, £50 per annum would be saved.

The minimisation of the use of paper and the identification of alternative disposal routes for waste paper might halve the number of bin collections, saving £8100 per annum for negligible cost.

The separating out of organic waste and reusable aggregate might also reduce the solid waste quantities by a further 25%, saving £4050. Then only five bins will be required, saving £500 per annum in rental charges.

The savings resulting from the more efficient use of metals are not precisely quantifiable. The annual bill for metals is around £10 million per annum. Assuming that 10% is wasted and that half of this might be saved by good practice, then £500 000 might be saved for little cost, except for the computer software needed for templating. The cost of this is of the order of tens of pounds.

WAS Step 33: Produce ranked investment portfolio

Table 8.28 presents the portfolio of *materials/waste* investment options for Green Products Ltd., ranked in order of rate of return on invested capital.

There are therefore opportunities to save £520 366 per annum for a negligible investment, only good practice.

WAS Step 34: Identify option conflicts

Whilst the various options for reducing paper waste quantities are in conflict, there are no conflicts in the above options.

WAS Step 35: Re-rank investment portfolio

This is not necessary.

Table 8.28 Materials/waste investment portfolio

Option	Description	Capital cost (£)	Savings per annum (£)	Straight rate of return (%)
A	Wait until tanks full	zero	50	infinity
B	Minimise paper waste	zero	8100	infinity
C	Organic waste processing	1	4050	infinity
		2	7666	infinity
D	Fewer bins	zero	500	infinity
E	More efficient use of metals	zero	500 000	infinity

Notes
(1) The cost for this is part of the research budget and so may be considered to be zero here.
(2) About 230 m^3, or 115 000 kg, of organic waste (0.5 kg per person) is produced each year. The calorific value of this waste is around 12 MJ/kg. Thus the waste contains 1 380 000 MJ, or 383 333 kWh. At a modest 2p/kWh, the waste as a fuel is worth at least £7667 per annum in a safe, storable and transportable form.

WAS Step 36: Develop waste management strategies

- Monthly monitoring and targeting
- Regular audits of waste
  - where does waste come from?
  - where does waste go?
- Maintain registers of:
  - legislation
  - environmental effects
  - incidents
- Education and training of employees

WAS Step 37: Establish five-year investment plan

The five-year investment plan for *materials/waste* retrofit measures is listed in Table 8.29 and the summary of savings is given in Table 8.30.

WAS Step 38: Set up monitoring and targeting and project management systems

WAS Step 39: Project manage

Back to the 39 steps of integrated environmental management . . .

IEM Step 26: Construct environmental accounts

This information is contained in EM Step 32, WAT Step 31 and WAS Steps 11 to 14.

Table 8.29 Five-year investment plan

Year	Option	Capital cost (£)	Accumulated savings (£)
1	A, B, C and D	0	20 366
2	E	0	540 732
3			1 061 098
4			1 581 465
5			2 101 831

Table 8.30a Materials summary (£)

Initial annual materials bill	11 000 000
Final annual materials bill	10 500 000 (95%)
Gross capital invested	zero
Net capital invested	zero
Accumulated savings	2 000 000

Table 8.30b Waste summary (£)

Initial annual waste disposal bill	17 150
Final annual waste disposal bill	4450 (26%)
Additional income from organic waste processing	7767
Gross capital invested	zero
Net capital invested	zero
Accumulated savings	101 830

IEM Step 27: Report on the environmental review

IEM Step 28: Identify and evaluate improvement options

Overall list of recommendations

- *Education.* Short seminars should be set up for management employees to highlight current laws and future directives, the environmental impacts of the company's activities, and to describe the operation of energy and environmental monitoring and targeting systems.

  Workshops on environmental issues for supervisory staff and shopfloor employees to foster responsibilities for departmental environments.
- *Environmental management*
  - Appoint an environmental manager.
  - Set up a monitoring and targeting system.
- *Energy*
  - Introduce a maximum demand monitoring and control system and institute maximum demand scheduling by coordinating the operation of large machines.

Table 8.31 Overall prioritised investment portfolio

Option	Description	Capital cost (£)	Savings per year (£)	Rate of return (%)
A	Compressed air	0	650	infinity
B	Lighting switch-off	0	10 000	infinity
C	Visual inspection of systems	0	1400	infinity
D	Washers on dripping taps	0	150	infinity
E	Wait until tanks full	0	50	infinity
F	Minimise paper waste	0	8100	infinity
G	Organic waste processing	0	11 717	infinity
H	Fewer bins	0	500	infinity
I	More efficient use of metals	0	500 000	infinity
J	Infra-red sensors at urinals	1500	16 000	1066
K	Lighting replacement	5000	10 000	200
L	MD control	20 000	20 000	100
M	Replace electrical heating	180 000	132 909	74
N	Ventilation control	136 674	80 000	58

- Investigate the use of electricity for weather-dependent heating purposes.
- Switch off unnecessary lights.
- Replace lights with low energy devices.
- Perform *input/output* analysis over the boiler system.
- Reduce compressed air pressure to 90 psi.
- Reschedule furnace operation to off-peak times.
- Investigate ventilating air system.
- Implement an energy management system.

- *Water*
  - Visual inspections of water distribution systems.
  - Repair leaks to water mains.
  - Fit new washers to dripping taps.
  - Install infra-red sensors at urinals.
- *Waste*
  - Improve the labelling of partially-used metal stocks
  - Pay attention to the storage and disposal of special wastes.
  - Minimise paper waste.
  - Seek cost-effective means of organic waste processing.
  - Seek the more efficient use of metals.

IEM Step 29: Produce a prioritised investment portfolio

This is provided in Table 8.31.

Option I will not be included in the following investment analyses as it is not directly environmentally related and also depends upon production procedures.

Thus 13 good investment options to save energy, water and materials with less than two-year paybacks have been identified. These cost in total £343 174, but result in annual financial savings of £320 363.

IEM Step 30: Construct an optimal overall investment schedule/project plan

The overall investment schedule and project plan is provided in Table 8.32 and the overall summary of savings is given in Table 8.33.

Note that the integrated environmental management approach allows a greater number of investment options to be adopted within the timespan of the project period.

Considering energy, water/effluent and waste separately results in final bills of 95% (energy), 73% (water/effluent) and 26% (waste) of the original bills, resulting in annual savings at the end of the project period of £70 900 and accumulated savings of £285 690.

When all three sectors are considered together, the final bills become 67% (energy), 73% (water/effluent) and 26% (waste) of the original bills, resulting in annual savings at the end of the project period of £320 363 and accumulated savings of £442 109.

IEM Step 31: Set the target cash flow forecast

IEM Step 32: Monitor the project

IEM Step 33: Audit the environmental management system

IEM Step 34: Make an internal environmental management audit report to top management

IEM Step 35: Internal reportage and dissemination

IEM Step 36: External reportage and dissemination

IEM Step 37: External validation of the environmental management system

IEM Step 38: Review the entire system and process

IEM Step 39: Go back to IEM Step 1 and start over . . .

Table 8.32 Overall investment schedule and project plan

Month	Option	Cost (£)	Accumulated savings (£)
1	A, B, C, D, E, F, G and H	0	2 714
2	J	1500	5 261
3			9 308
4	K	5000	9 188
5			14 068
6			18 948
7			23 861
8	L	20 000	10 325
9			16 922
10			23 469
11			30 016
12			36 563
13			43 110
14			49 657
15			56 204
16			62 757
17			69 298
18			75 845
19			82 392
20			88 939
21			95 426
22			102 033
23			108 580
24			115 127
25			121 674
26			128 221
27			134 768
28			141 315
29			147 862
30			154 409
31			160 950
32			167 503
33			174 050
34			180 597
35	M	180 000	18 220
36			35 843
37			53 466
38			71 089
39			88 712
40			106 735
41			123 958
42			141 581
43	N	136 674	29 196
44			53 485
45			77 724
46			102 063
47			126 352
48			150 641
49			442 109

Table 8.33 Overall summary of savings

Initial annual bill (energy, water, effluent, waste)	£846 324
Final bill	£525 961 (64%)
Gross capital invested	£343 174
Net capital invested	zero
Accumulated savings	£442 109

Other studies

The foregoing analysis has identified potential savings of over £300 000/annum for Green Products Ltd. Of the investment projects formulated, 10% have zero financial cost, and the remaining 90% have paybacks of less than one year.

Cranfield University

The integrated environmental management procedure demonstrated here has been developed considerably at Cranfield University during recent years. Its procedure has been successfully applied to a metals refinery, a paper mill, a sugar refinery, a tannery, a waste water treatment works, plants for the manufacture of large power components, industrial gases, and car manufacture, an airways catering centre, a high street retailer, a research and development centre, various hospitals and the counties of Bedfordshire and Cornwall.

A summary of results from some of these projects is provided in Table 8.34.

The organisations investigated had total electricity bills of £5.0 million, a total fuel bill of £4.7 million, total water and effluent bills of £1.2 million and a total waste disposal bill of £0.25 million, making an overall total utilities bill of just over £11.0 million.

Sixty-six good cost-effective recommendations for system improvement options were identified. These cost in total £1.15 million and resulted in predicted savings of £2.4 million per annum for the sites concerned—an overall payback period of less than six months!

Aire and Calder (3)

The March Consulting Group in the UK conducted a study on reducing costs and improving performance through waste minimisation. Eleven participating companies were involved. Potential savings totalling £2 million were identified. Of these, 10% of investment projects had zero cost, 60 had financial payback periods of less than one year and only 10% of all the projects had payback periods greater than two years.

Table 8.34a Summary data from a number of integrated environmental management audits: initial bills (£)

Company	Annual maximum demand bill	Annual electricity bill	Annual fuel bill	Annual water bill	Annual effluent bill	Annual waste disposal bill
A	28 777	127 732	99 869	9 403	120 000	60 000
B			132 199			
C		1 500 000	1 900 000	750 000		
D		75 000	7 500	2 045	5 700	
E		480 000	240 000	144 000	76 000	185 000
F	50 000	260 000	240 000	60 000		
G	145 000	915 000	316 666		12 000	
H		1 200 000	1 800 000			

Notes

Total electricity bill £5.0 million
Total fuel bill £4.7 million
Total water/effluent bill £1.2 million
Total waste disposal bill £0.25 million
Total utilities bill £11.0 million
Not all the figures for effluent and waste disposal were available.

Table 8.34b Summary data from a number of integrated environmental management audits: recommendations

Recommendation	Saving	Cost (£)	Annual saving (£)	SPBP (straight payback period) (yrs)
Reduce electrical supply capacity	money	0	4000	0
Reduce inside air temperatures	electricity	0	150	0
Eradicate unnecessary lighting	electricity	0	6000	0
Reduce inside air temperatures	coal	0	800	0
Control of machinery	electricity	0	11 000	0
Eliminate electrical heating	electricity	0	1300	0
Attention to water use	water/effluent	0	1300	0
Tariff change	money	0	1300	0
Reduce inside air temperatures	gas	0	2800	0
Control of heaters	gas	0	5000	0
Reduce electrical supply capacity	money	0	2400	0
Switch off unnecessary lights	electricity	0	10 000	0
Reduce compressed air pressure	electricity	0	650	0
Waste reduction	materials/ disposal costs	0	22 000	0

continued

Table 8.34b *(continued)*

Recommendation	Saving	Cost (£)	Annual saving (£)	SPBP (straight payback period) (yrs)
Reduce inside air temperatures	gas	600	250 000	0.002
Valve flow adjustments	water	10	1500	0.007
Process change	gas	10 000	315 000	0.03
Repair water leaks	water	100	3000	0.03
Maximum demand control	money	1000	25 000	0.04
Maintenance of steam systems	gas	5000	130 000	0.04
Electrical load control	electricity	10 000	225 000	0.04
Ventilation control	coal	1000	22 000	0.05
Recycle effluent	water/effluent	1000	14 000	0.07
Reduce water temperatures	gas	300	3100	0.1
Control of space heating	gas	6000	60 000	0.1
Control of compressed air system	electricity	3200	32 000	0.1
Install cistern dams	water	70	700	0.1
Control of steam boiler	gas	500	38 000	0.13
Fit control valves	gas	500	3500	0.14
Maintenance of compressed air systems	electricity	5000	36 000	0.14
Steam metering	gas	1250	8500	0.15
High temperature insulation	gas	3600	23 000	0.16
Insulation of heaters	gas	550	2900	0.19
Use of effluent	gas	10 000	50 000	0.2
Fit dimmer controls to lights	electricity	5000	25 000	0.2
Use of waste heat	gas	50 000	222 000	0.23
Draughtproofing	gas	250	1000	0.25
Air preheat with exhaust gases	gas	40 000	145 000	0.28
Combustion of solvent	coal	1000	3600	0.3
Install flush regulators	water	300	1000	0.3
Furnace baffles	fuel	8000	23 000	0.35
Insulation of furnaces	fuel	8000	20 000	0.4
Reduce condensate losses	gas	3000	7400	0.41
Sparge pipe for steam blowdown	gas	500	1200	0.42
Replace steam boiler	gas	13 000	30 000	0.43
Waste heat recovery	fuel	38 000	80 000	0.48
Monitoring and targeting	gas/electricity	4300	7600	0.57
Install tap flow regulators	water	70	120	0.58
Install urinal flush controllers	water	270	450	0.6
More efficient furnaces	fuel	180 000	230 000	0.78
Infra-red detectors in urinals	water	1100	1300	0.85
Combustion air preheat	gas	1880	2100	0.89
High temperature insulation	gas	29 000	32 000	0.9
Replace lights	electricity	5000	5000	1.0
Furnace control	fuel	31 000	25 000	1.24

continued overleaf

Table 8.34b *(continued)*

Recommendation	**Saving**	**Cost (£)**	**Annual saving (£)**	**SPBP (straight payback period) (yrs)**
Replacing lights	electricity	1500	1000	1.5
Use of compressor waste heat	coal	10 000	6000	1.7
Tap regulators	water	300	140	2.14
Reduce heater exit temperatures	gas	12 000	5000	2.40
Abolish electrical heating	money	180 000	68 000	2.64
Process improvements	materials	450 000	167 000	2.70
Lighting controls	electricity	1500	550	2.72
Maintenance of heaters	gas	15 000	4500	3.33

Notes
Total annual saving £2.4 million
Total capital cost £1.15 million
Overall payback period less than six months!

Project Catalyst (4)

This UK project involved 14 participating companies and sought opportunities for reducing emissions and discharges to the three environmental media. Potential financial savings totalling £8.9 million per annum were identified. Of these, almost £2.5 million of savings involved zero cost and almost £3 million had a payback of less than one year.

REFERENCES

(1) P.W. O'Callaghan, *Energy Management*, McGraw-Hill, Oxford, 1993.
(2) *CIBSE Guide*, Chartered Institute of Building Services Engineers, London, 1993.
(3) March Consulting Group, *Reducing Costs and Improving Performance through Waste Minimisation—'The Aire and Calder Experience'*, Version 1.2, Manchester, 1994.
(4) W.S. Atkins/March Consulting Group/Aspects International, *Project Catalyst, Report to the Demos Project Event*, Manchester Airport, 27 June 1994.

9 Summary

This chapter reviews the contents of this book and points to some future challenges faced by companies.

9.1 SUMMARY

Chapter 1 reviewed and discussed global and local environmental problems.

Global environmental problems include the greenhouse effect leading to global warming and climate changes, ozone layer depletion leading to increased ultra-violet radiation reaching the surface of the earth, and water pollution and eutrophication due to excess nitrate run-off.

Local environmental problems include traffic pollution from road vehicles, roads and infrastructure, air pollution from vehicles causing respiratory problems and other health hazards, air pollution from power generation and factories, water pollution from people, industry and agriculture and land pollution from increasing quantities of domestic and industrial waste.

International reactions to these growing problems include the Toronto and Montreal Protocols, which sought to reduce emissions of carbon dioxide and CFCs respectively. Within the European Union, numerous environmental Directives have been issued. Because the Commission has little direct control over national legislations, individual member states are expected to act according to these directives within the legislative frameworks of the individual states under the terms of subsidiarity. Five European Commission Environmental Action Programmes have so far been established to address pollution issues.

In the UK, Her Majesty's Inspectorate of Pollution (HMIP) and the National Rivers Authority were established and new legislation and regulations followed, culminating in the 1990 Environmental Protection Act, which deals simultaneously with emissions to the three environmental media: air, water and land. The Act introduced the concept of integrated pollution control, from which the concept of integrated environmental management emerged. The Environmental Protection (Prescribed Processes and Substances) Regulations 1991 lists the processes for which an authorisation from HMIP is required. One objective is that the best available techniques not entailing excessive costs (BATNEEC) are used to prevent or minimise pollutant releases to the environment.

The British Standards Institution (BSI) produced a specification for environmental management systems (BS 7750) and the European Union introduced the Eco-management and Audit regulation.

As a result, companies now have an urgent need to:

- comply with new environmental legislation and so avoid fines and other penalties
- conserve resources by using use less input materials and energy and so save money
- adopt high environmental standards by obtaining BS 7750 accreditation, and/or to comply with the requirements of the EU Eco-management and Audit Scheme.

There is no doubt that the costs of energy, electricity, water supplies and raw materials will rise substantially in real terms, as the supply companies comply with new environmental legislation and standards. The use of less input resources results in less emissions to air and water and less solid waste. This leads to lower levels of pollution taxes, effluent charges and waste disposal costs. These costs will also rise substantially over the coming years as the effluent and waste disposal operators also have to comply with high environmental standards and legislation.

Integrated environmental management is defined as the integration of environmental management with environmental engineering, covering global and local problems, legislation, standards, economics, policy and impact with respect to energy and air pollution, water and effluents, the use of materials, the production of waste, the production of land pollution, noise abatement, products and ecolabelling. It was explained that it is necessary to conserve resources and so save money whilst protecting the environment, to avoid legal action, fines and pollution taxes and to be seen to be environmentally responsible to gain a competitive advantage. The application of IEM leads to effective integrated pollution control.

Chapter 2 dealt with EC and UK legislation with respect to environmental pollution. It reviewed the concepts of nuisance and duty of care in this regard and then examined environmentally related EC Action Programmes and Directives. The major EC Directives, UK Acts and Regulations concerning air pollution, water pollution and waste on land were listed chronologically. The UK Environmental Protection Act 1990 was considered in some detail. The concept and aims of integrated pollution control were defined and the principle of BATNEEC described. An Appendix to this chapter contains the list of prescribed processes and substances and examples of BATNEEC for the reduction of emissions are collated. The application procedure for HMIP IPC Authorisation to emit is also included. Data taken from dozens of IPC process guidance notes were assembled into a list of general BATNEEC.

Chapter 3 reviewed the European Union's Eco-management and Audit Scheme (EMAS) which allows voluntary participation by companies, so that they may assess their environmental impacts, set policy and plan for improvements. The concepts of environmental policies, environmental objectives, environmental programmes, environmental management systems, environmental reviews, environmental auditing, environmental statements and environmental information disclosure were introduced. The essential components of the British Standards Institution's specification for environmental management systems, BS 7750:1992 were listed, which is applicable to any organisation which wishes to assure itself of compliance with a stated environmental policy, and to demonstrate such compliance to others. The voluntary EC Environment Council Ecolabelling Regulation 1992, intended to promote the design, production, marketing and use of products which have a reduced environmental impact during their entire life cycle and to provide consumers with better information on the environmental impact of products, was examined and its implications discussed. The concepts of product environmental audits, cradle-to-grave environmental impacts of products, life cycle analysis, and life cycle assessment were introduced.

Chapter 4 showed how integrated environmental management integrates the requirements of environmental legislation, the EC Eco-audit Regulation, BS 7750, the need to conserve energy, water and materials, the need to reduce polluting emissions to air, effluents to water and solid wastes to land, and the need to save money in the process, via reduced resource and utility costs, reduced effluent and waste disposal charges, reduced pollution taxes, the avoidance of fines and legal penalties and the avoidance of clean-up costs. It was shown that the systematic application of integrated environmental management allows all necessary retrofits and modifications to technical activities to be financially resourced from within the environmental management project itself. The cyclic chain of activities to be conducted involves: education, commitment, an initial review, policy, organisation, compliance, a scoping study, management, documentation, a full review, the identification of options leading to recommendations, audits and reportage, concerning energy and emissions to air, water and effluents, waste on land, materials, products and packaging, transport and noise. The 39 sequential steps of integrated environmental management were listed.

Chapter 5 dealt with fossil fuel combustion and the concomitant polluting emissions to air. Trends in pollutant emissions were examined for the European Union and the UK. For the latter, the rapidly accelerating growth in polluting emissions from road traffic was identified as a major problem of immediate concern. The types and calorific values of the various fuels were reviewed and example calculations of energy release and pollutants produced provided. Both energy-related and non-energy-related emissions to air were considered. Rules for the efficient and cost-effective conservation of energy were included, as well as questions for the client interview and a checklist for energy managers. The 39 steps to optimise energy management investments at a site were constructed. The procedures of energy management were demonstrated via a worked example. It was shown that a site can simultaneously save energy, save money, reduce pollution and so protect the environment.

Chapter 6 dealt with water supplies, effluents and water management. The problems arising from increased water usage (e.g. supply shortages, effluent and sewage disposal, pollution and eutrophication of water courses) were discussed and the appropriate legislation relating to water reviewed. Parameters to consider when assessing water quality were listed and water treatment processes described. The uses for water were examined and a procedure for water accounting was developed. The 39 steps to optimise water management investments at a site were constructed. Rules for the efficient and cost-effective conservation of water were included, as well as questions for the client interview and a checklist for water managers.

Chapter 7 dealt with waste management. The problems of resource depletion and pollution arising from consumerism, the throwaway society and increased rates of waste generation were reviewed. Waste arisings were classified into inert wastes, semi-inert wastes, putrescible wastes, difficult wastes, special wastes and prohibited wastes. The various methods of waste disposal were evaluated and compared. The procedures involved in materials auditing, as well as rules for the efficient conservation of materials, were developed. Questions for the client interview and checklists for waste managers were listed. The 39 steps to optimise waste management investments at a site were constructed. Techniques of materials conservation (e.g. reclaim, recycle, reuse, energy from waste, waste-derived fuels) leading to cost-effective waste minimisation were reviewed and evaluated.

Chapter 8 contained a fully worked integrated environmental management case study of a corporate site. The 39 steps of environmental management, energy management, water management and waste management were worked through systematically. Thirteen good investment options to save energy, water and materials were identified. Eight of these options involved zero capital cost and the remaining five options had financial payback periods of less than two years. By adopting a no-risk investment plan, where perceived and realised savings are reinvested in further cost-effective options, annual financial savings of £320 363 were seen to be possible for the site for no net capital investment. In addition to this, the project plan indicated that a sum of £442 109 of accumulated savings plus interest will have accrued at the end of the five-year project period. This case study demonstrated the importance of the integrated environmental management approach, as it allows more investment options to be adopted earlier in the project, resulting in maximising financial savings within the duration of the project. The summarised results of similar studies, involving 34 companies in total, showed that similar savings are possible for a range of industrial and commercial sectors. It was demonstrated, therefore, that it is possible for a company simultaneously to comply with environmental legislation, adopt high environmental standards, save resources, save money and protect the environment.

9.2 THE FUTURE

Resources are used and pollution produced in the exploration, extraction, supply, conversion and utilisation of fossil fuels and raw materials, in transportation, and in the manufacture, advertising, marketing, distribution, retailing and utilisation and disposal of products. Conservation-conscious planning and design can reduce energy and materials consumption and concomitant environmental pollution in each of these activities.

Future environmental auditing will adopt a 'cradle-to-grave' approach. Environmental ratings and comparisons among products will consider the energy and raw materials requirements and the air, water and soil pollution produced by a product in production, distribution, utilisation and ultimate disposal to the environment.

Toxic materials, unnecessary packaging, disposable products, unnecessary products, dangerous products, irreparable items, energy-intensive products and systems, and environmentally polluting items (in manufacture, use or disposal) will be avoided.

The concept of sustainable development was central to the UN Earth Summit held at Rio de Janeiro in June 1992. Agenda 21 emerged from this conference, calling for governments to prepare national strategies for sustainable development.

The Fifth European Commission Environmental Action Programme focused upon the improved enforcement of existing legislation, but also indicated the probable development of EU environmental policies leading to further new legislation over coming years. This policy had been aimed at controlling and reducing the environmental impact of industry (i.e. end-of-pipe clean-up). The next steps will involve the adoption of clean technologies (e.g. with regard to packaging, landfills, incinerators, etc.) and the development of environmentally sound products. In other words, EC directives, national legislation and standards will start to creep up the disposal and production pipes.

The consumer will be told:

- *'Thou shalt not discard . . .'*
- *'Thou shalt not accept products which . . .'*

The producer will be told:

- *'Thou shalt not use . . .'*
- *'Thou shalt not produce . . .'*

Products will be built to last, built simple and built in modular construction for ease of repair.

There is no doubt that, in order to reverse growing global and local environmental problems, a sustained effort must be made to reduce substantially the quantities of pollutant emissions to air, water and land.

The widespread application of sensible and systematic integrated environmental management procedures and practices, as outlined in this book, will allow the necessary pollution reductions to be made in the short term. In the medium term, consumer attitudes towards consumerism versus conservationism will swing to favour environmental protection.

In the long term, social changes with respect to transportation systems and the work ethic will ensure renewable resource economies, where nations, having very much reduced demands for energy and material resources, are sustained by renewable energy technologies: biomass, solar, wind, waves, tides and hydroelectric power.

Technical annex: summary of fundamental concepts, relationships and equations

This Annex contains fundamental information regarding the technical aspects of integrated environmental management.

Contents

A1.1 Cycles of life
- The carbon cycle
- Photosynthesis
- Respiration
- Decay
- Sea cycles
- The nitrogen cycle
- The phosphorus cycle
- Animals and vegetation

A1.2 Energy equations
A1.3 The U-value
A1.4 Radiation involving gases
A1.5 Boilers
- Energy conservation in boilers

A1.6 Building heat balance
A1.7 Thermal comfort
A1.8 Heat gains to buildings
- Solar gains

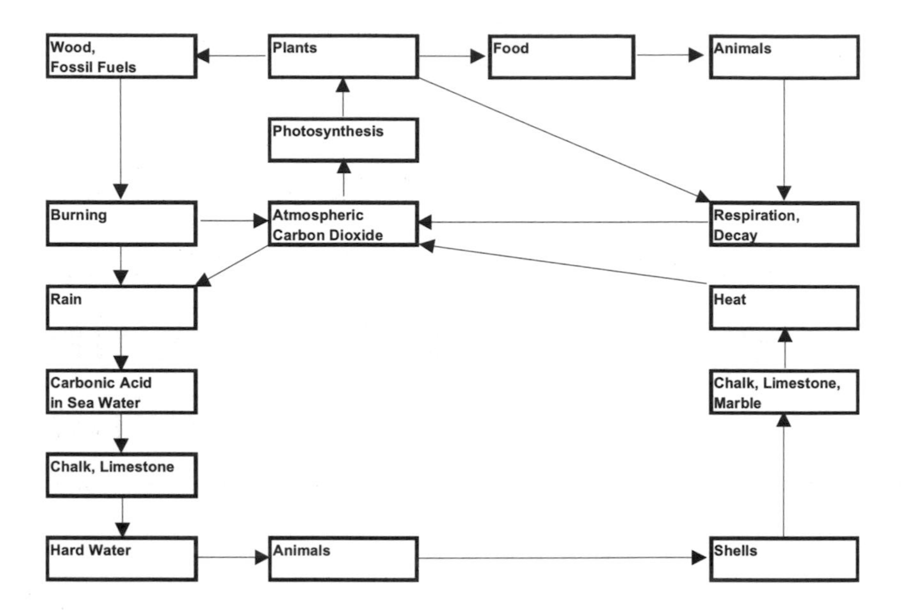

Figure A1.1 The carbon cycle

Heat gains from people
Heat gains from electrical devices
A1.9 Carbon taxes

A1.1 CYCLES OF LIFE

The carbon cycle

The carbon cycle is the circulatory chain of events by which carbon is exchanged between animals and plants and the global environment. Figure A1.1 depicts the carbon cycle on Earth. The diagram is split into two halves. The upper half describes the terrestrial exchanges of carbon from the atmosphere to plants and animals. The lower half shows how carbon is 'fixed' in the oceanographic environment.

Photosynthesis

Atmospheric carbon dioxide is absorbed by plants via the process of photosynthesis, the process by which carbohydrates are formed in living plants by the action of sunlight on chlorophyll. This absorption occurs directly by absorption from the air through the underside of the leaves, or indirectly from water via the roots. Oxygen is formed and released to the atmosphere in the process. The chemical equation governing the process is as follows:

Carbon dioxide + water + energy from sunlight = carbohydrate + oxygen

or, in symbolic notation,

$$6\ CO_2 + 6\ H_2O + \text{solar energy} = C_6H_{12}O_6 + 6\ O_2$$

Now, the atomic weights of carbon, C, oxygen, O, and hydrogen, H, are as follows:

C 12
O 16
H 1

The molecular masses (grams/mole or kg/kmole) of carbon dioxide, water, carbohydrate and oxygen are therefore:

Substance	Chemical formula	Molecular mass
Carbon dioxide	CO_2	44
Water	H_2O	18
Carbohydrate	$C_6H_{12}O_6$	180
Oxygen	O_2	32

Thus:

6×44kg of carbon dioxide + 6×18 kg of water + sunlight energy produce 180 kg of carbohydrate + 6×32 kg of oxygen

or

264 kg of carbon dioxide + 108 kg of water + sunlight energy produce 180 kg of carbohydrate + 192 kg of oxygen

or

1.47 kg of carbon dioxide + 0.6 kg of water + sunlight energy produce 1 kg of carbohydrate + 1.067 kg of oxygen

Furthermore, 1 kg of carbohydrate provides 16.9 MJ of energy when burnt or eaten and metabolised. Thus, from the conservation of energy, the net amount of sunlight energy required to produce 1 kg of carbohydrate is 16.9 MJ.

Therefore:

1.47 kg of carbon dioxide + 0.6 kg of water + 16.9 MJ of sunlight energy produce 1 kg of carbohydrate + 1.067 kg of oxygen.

It has been estimated that tropical rainforest growth may fix 1–2 kg of carbon per m^2 per year, but that areas such as barren tundra will fix less than 1% of this amount.

Respiration

Respiration by animals is the process by which oxygen is transported from the lungs to the muscles by the haemoglobin in the blood. The carbohydrates oxidise to form carbon dioxide which is transported back to the lungs to be expelled to the atmosphere.

Respiration is thus the opposite process to photosynthesis, carbohydrates being converted to carbon dioxide, water and chemical energy by living creatures. Its chemical reaction is described by:

Carbohydrate + oxygen = carbon dioxide + water + muscular energy

or, in symbolic notation,

$$C_6H_{12}O_6 + 6\ O_2 = 6\ CO_2 + 6\ H_2O + \text{muscular energy}$$

Thus 180 kg of carbohydrate needs 192 kg of oxygen to produce 264 kg of carbon dioxide and 108 kg of water providing muscular energy.

Or, 1 kg of carbohydrate needs 1.067 kg of oxygen to produce 1.47 kg of carbon dioxide and 0.6 kg of water providing 16.9 MJ of muscular energy.

The carbohydrates in the body emanate from food, which originates directly, or indirectly, from plants.

Plants are the only living things able to manufacture their own food internally, using photosynthesis to convert water, carbon dioxide and minerals into sugars, proteins and other energy sources. During photosynthesis, energy from the sun is converted into food energy in the form of glucose. To liberate this energy, this glucose reacts with oxygen to form carbon dioxide and water.

Thus plants also respire continuously. The chemical reaction involved is described by the respiration equation above, so oxygen is absorbed and carbon dioxide is released. At night, in the absence of sunlight, there is no process of photosynthesis and carbon dioxide is released.

Some living organisms respire without using oxygen. The process involved is that of anaerobic respiration (e.g. yeast in the process of fermentation).

Decay

The carbon 'fixed' by photosynthesis in plants and animals is eventually returned to the atmosphere as carbon dioxide when plants and animals die and the dead organic matter is consumed by the decomposer mechanisms. The overall chemical reaction involved is described by the respiration equation above.

Sea cycles

The lower half of Figure A1.1 shows how carbon is exchanged between the atmosphere and the oceans. Atmospheric carbon dioxide is washed down by rain to form carbonic acid, H_2CO_3, in sea water. Phytoplankton are single-celled organisms that take up carbon dioxide from sea water and, by photosynthesis, use it to produce carbohydrates and oxygen, which dissolves in the water. Zooplankton and fish consume the carbon fixed by the phytoplankton. When these predators and uneaten phytoplankton die, the carbon is returned as carbon dioxide dissolved in seawater. Shellfish fix carbon as carbonates in their shells, which are eventually discarded to the ocean floor to form chalk, limestone and marble sediments. If these substances are heated, carbon dioxide is released to the atmosphere, completing the oceanographic carbon cycle.

Table A1.1 lists solar conversion efficiencies and carbon fixation rates for various crops.

The nitrogen cycle

Figure A1.2 shows the processes involved in the nitrogen cycle.

Atmospheric nitrogen may be absorbed by plants in two principal ways:

(1) Lightning can oxidise atmospheric nitrogen to form dilute nitric acid with water. Rain washes this acid to the soil where nitrates are formed. Bacterial and chemical actions in the soil convert these into soluble nitrates which are taken up by the roots of plants.
(2) Leguminous plants have nodules on their roots which contain colonies of bacteria capable of converting atmospheric nitrogen directly into soluble nitrogen compounds.

A secondary cycle involves the animals, which consume plants as food and reject manure and urine. These waste products enter the soil where they are converted by bacterial and chemical actions to soluble nitrates to be taken up by the roots of plants. When plants and animals die and decay, further nitrates are formed in the soil.

In aquatic environments, nitrogen is fixed by blue-green algae.

Denitrifying bacteria in the soil decompose oxides of nitrogen to nitrogen to be returned to the atmosphere, completing the cycle.

The phosphorus cycle

The phosphorus cycle operates as shown in Figure A1.3.

Phosphate in rocks is leached to soil and water and soluble phosphates are taken up by plant roots. The plants are eaten by animals, which eventually die and decay to return the phosphates to the soil.

The phosphorus cycle is a sedimentary cycle, the exchanges involving the action of water, as do the cycles for calcium, iron, potassium, manganese, sodium and

Table A1.1 Solar conversion efficiencies and carbon fixation rates for various crops

Plant type	Location	Total radiation falling on location ($MJ\ m^{-2}\ yr^{-1}$)	Calorific value ($MJ\ kg^{-1}$)	Dry yield ($kg\ m^{-2}\ yr^{-1}$)	Carbon dioxide fixed per annum ($kgCO_2m^{-2}\ yr^{-1}$)
Alfalfa	Average US	5270	15.1	0.64	0.94
Southern pine	South US	6050	16.3	0.45–1.12	0.66–1.65
Oak-pine forest	New York	4181	16.3	1.21	1.78
Reed canary grass	Midwest US	5270	15.1	1.42	1.66
Hybrid poplar	Pennsylvania	4930	13.1	0.90–1.80	1.32–2.65
Bermuda grass	Alabama	6050	13.1	1.80–2.46	2.65–3.62
Sycamore	Georgia	6050	13.5	0.36–2.51	0.53–3.69
Cattail swamp	Minnesota	4260	15.1	2.51	3.69
Marine algae	Nova Scotia	4810	15.1	2.06–2.62	3.03–3.85
Corn	Average US	5270	15.1	2.51–4.01	3.69–5.89
Sugar cane	Louisiana	6050	15.1	4.48	6.59
Sewage pond	California	6310	15.1	5.62	8.26

Notes
1 kg of carbohydrate absorbs 1.47 kg of carbon dioxide (0.27 kg of carbon).

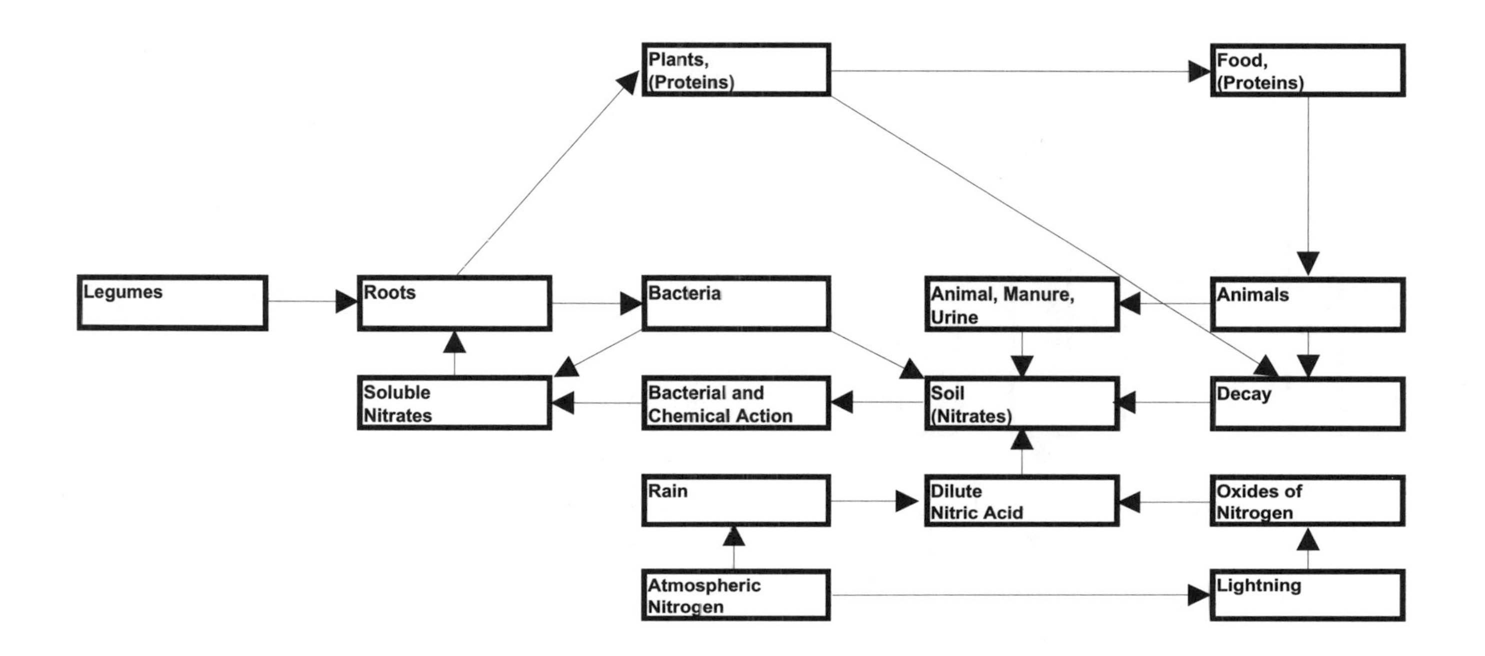

Figure A1.2 The nitrogen cycle

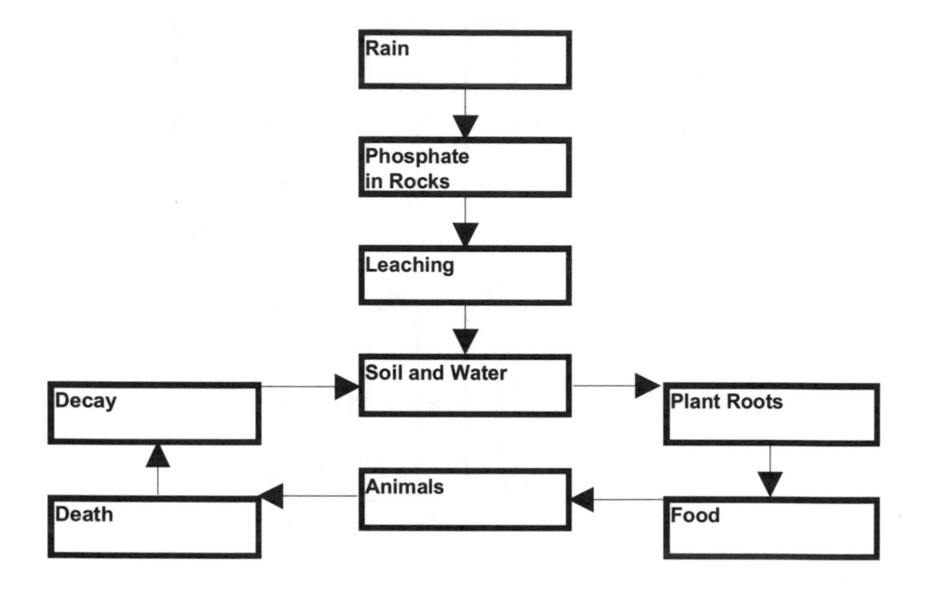

Figure A1.3 The phosphorus cycle

sulphur, although the latter cycle also involves the oxidation of atmospheric sulphur dioxide to sulphur trioxide, which combines with rain water to form sulphuric acid. This reacts with other elements in the soil to form sulphates. Bacterial action converts the sulphates into hydrogen sulphide, which is oxidised to sulphur dioxide and returned to the atmosphere.

Animals and vegetation

For life to exist on Earth, the cycles of life must continue to operate.

Vegetation takes in oxygen and gives out carbon dioxide in the process of photosynthesis, whilst animals take in oxygen and give out carbon dioxide in the process of respiration. The carbohydrates in the body emanate from food, which originates directly, or indirectly, from plants. Atmospheric nitrogen enters living organisms, being absorbed into green plants in the form of nitrates, the plants being eaten by animals to build proteins. Nitrogen is returned to the ecosystem through the animal's excreta or via decay. Phosphorus is needed by plants to produce good root systems. The plants are eaten by animals, producing bones and nerve tissue. When animals die and decay, phosphorus is returned to the soil. Cycles for calcium, iron, potassium, manganese, sodium and sulphur involve similar exchanges between the plants and the animals.

Animals cannot exist without vegetation, although vegetation could exist without animals.

A1.2 ENERGY EQUATIONS

Energy (W) may be contained in a flow system as:

potential energy	mgh
kinetic energy	$mu^2/2$
pressure energy	mP/ρ
heat energy	$m\ c_v\ T$

where

m is the mass flow (kg s^{-1})
g is the acceleration due to gravity (= 9.81 m s^{-2})
h is height (m)
u is the mean velocity (m s^{-1})
P is the absolute pressure (N m^{-2})
ρ is density (kg m^{-3})
c_v is the specific heat at constant volume (J $kg^{-1}K^{-1}$)
and T is the absolute temperature (K)

Note that 1 Newton is the SI unit of force which gives a mass of 1 kg an acceleration of 1 m s^{-2}. It thus has the units of kg m s^{-2}.

1 J = 1 Nm and so has the alternative units kg m^2s^{-2}.
1 W = 1 Nm s^{-1} and so has the alternative units kg m^2 s^{-3}.

A1.3 THE U-VALUE

In the majority of situations, heat is transferred from one fluid to another across a solid wall. This process is known as recuperative heat transfer. It is customary to employ an overall heat transfer coefficient, U, based upon the overall temperature difference between the two fluids, ΔT.

The heat transferred between the fluids is then given by:

$$Q = UA\Delta T$$

Table A1.2 lists measured U-values for some commonly encountered structures.

Table A1.2 Common U-values (W $m^{-2}K^{-1}$)

Component	U-value
105 mm solid brickwork	3.3
220 mm solid brickwork	2.3
335 mm solid brickwork	1.7
260 mm cavity brickwork	0.8–1.5
Corrugated sheeting	5.3[1]
150 mm solid concrete	3.4
6 mm single glazing	5.6[1]
2 mm airspace double glazing	2.9
3 mm airspace double glazing	4.0
Uninsulated flat roof	≈3.0
Uninsulated pitched flat roof (35° slope)	≈1.5
Solid floor	≈0.3[2]

Notes

(1) These values are heavily dependent upon the values of filmside heat transfer coefficients adopted. The solid U-values, i.e. neglecting boundary resistances, of 5 mm metal sheeting and 6 mm glazing are respectively ≈ 9000 W $m^{-2}K^{-1}$ and ≈ 130 W $m^{-2}K^{-1}$.

(2) Referred to inside minus outside air temperatures.

Source: *CIBSE Guide*, Chartered Institute of Building Services Engineers, London, 1993.

A1.4 RADIATION INVOLVING GASES

Many common gases, such as oxygen, nitrogen, hydrogen and dry air, are practically transparent to thermal radiation. Carbon dioxide, water vapour, sulphur dioxide, carbon monoxide, ammonia, hydrocarbon and alcohol vapours emit and absorb radiation only between narrow bands.

Table A1.3 lists emission and absorption bands for common gases and vapours.

The greenhouse gases, carbon dioxide, CO_2, nitrous oxide, N_2O, the CFCs, methane, CH_4, tropospheric ozone, O_3 and water vapour, H_2O, all impede the transmission of long-wavelength radiation from the Earth.

Stratospheric ozone inhibits the transmission of short-wavelength ultra-violet radiation from the sun.

A1.5 BOILERS

Boilers are usually used for steam raising or for heating hot water for space heating, to provide domestic hot water, or for other industrial purposes. Steam is used for heating, cleaning, autoclaving and for driving turbines to provide electrical power.

Table A1.3 Emission and absorption bands for common gases and vapours

Substance	Wavelength band λ (μm)
Carbon dioxide	2.36–3.02
	4.01–4.80
	12.5–16.5
Water vapour	2.24–3.27
	4.80–8.50
	12.00–25.00

Values for efficiency calculated from analysing stack losses must be further discounted for losses due to casing losses and blowdown losses. These are typically of the order of 7% for a steam boiler.

Radiation, convection and miscellaneous losses are approximately as follows:

Type of boiler	% loss at MCR (maximum continuous rating)
Modern package	3.0
Water tube	3.5
Economic wet-back	4.0
Economic dry-back	5.0
Sectional boiler	5.0
Cornish and Lancashire	6.5
Vertical boiler	6.5
Sectional boiler with no heat transfer under the furnace	8.5

For a boiler operating at loads below the maximum continuous rating, the above losses may be multiplied by the turndown ratio based upon metered fuel supplied. The blowdown loss is taken as zero for hot water boilers.

Table A1.4 is a pro forma for a boiler efficiency report.

Table A1.5 lists typical efficiencies for different boiler types.

Energy conservation in boilers

Heat losses from the flue gases can be minimised by running boilers efficiently. The amount of combustion air should be limited to that necessary to ensure complete combustion of fuels at all times, with a slight margin of excess air to suit the installation. Excess air increases flue losses. Too little air results in incomplete combustion of the fuel. Heat transfer surfaces should be kept clean and soot formation should be avoided at all times.

Table A1.4 Pro forma for boiler efficiency report

Type of boiler	**Hot water/steam/other**
Make	
Maximum continuous rating, MCR (kW)	
Working pressure (N m^{-2})	
Working temperature (K)	
Fuel type and grade	
Type and make of stoker/burner	
Test data	
Loading, % MCR	
Fuel	
Chemical analysis	
Gross calorific value (MJ kg^{-1})	
Flue gas exit temperature (K)	
Boiler house temperature (K)	
CO_2 in flue gas, %	
O_2 in flue gas, %	
CO concentration in flue gas	
Analyses	
Sensible heat of dry flue gas %	
Enthalpy of water vapour in flue gas %	
Unburnt gas (CO) losses %	
Radiation, convection and miscellaneous %	
Blowdown losses %	
Total losses %	
Efficiency (100—total losses)	
Estimate of annual plant efficiency %	

Table A1.5 Typical efficiencies (%) for different boiler types (based on gross calorific value)

Boiler type	**Fuel**	**Efficiency (%)**
Industrial water tube boiler	Oil	84–87
with economiser	Gas	80–83
Shell boiler	Oil	81–84
Shell boiler without	Gas	77–80
economiser	Coal	77–81
Sectional boilers	Oil	75–80
Low temperature hot water	Gas	71–76
	Coal	71–76

If the combustion mixture is set correctly and heat transfer is uninhibited, the addition of fuel additives and other 'fuel-saving' devices are unlikely to produce further improvements.

An *economiser* can save up to 5% of the fuel used by using waste flue gases to preheat the feed water via a recuperative heat exchanger.

The introduction of *baffles* to bring combustion gases into closer contact with water or steam tubes is likely to be effective only on old boilers.

A recent development sprays water through the flue gases to recuperate some heat for use elsewhere.

A *rotary regenerator* may be used to recover some of the heat in exhaust gases to heat air for space heating or for preheating the combustion air.

Flue dampers might be used to close down flues when boilers sharing the same chimney close down intermittently.

Radiation and convection losses may be reduced by proper insulating techniques. These become appreciable especially under low loads.

Boiler blowdown is necessary to remove sludge from precipitated salts, and so prevent scale forming as an extra thermal resistance on the tubes. Frequencies and durations of blowdown should be just enough to achieve this—and no more. A heat exchanger might be installed to recover heat from blowdown steam or water, to preheat feed water.

Correct chemical water treatment is essential to prevent scale formation and to reduce or eliminate corrosion in the boiler. In steam systems, as much condensate as possible should be recovered and returned to the boilerhouse to gain heat, make-up water and treatment chemicals.

Continuous logs of boiler performance should be kept.

The *Fuel Efficiency Booklets 14 (Oil)*, *15 (Gas)* and *17 (Coal)*, produced by the UK Energy Efficiency Office, contain some useful charts for estimating boiler efficiencies rapidly.

A1.6 BUILDING HEAT BALANCE

Energy enters a building in fuels, electricity, as sundry gains from people and as solar gains. Heat energy leaves the system boundary as stack losses, fabric transmission losses and ventilation losses. The annual heat balance equation is as follows:

Energy supplied in fuels and electricity
+ sundry gains

= fabric transmission heat losses
+ ventilation losses
+ stack losses

$$E + SG$$

$$= (UA + nV/3)(T_i - T_o) \times (\text{number of hours/annum})/1000 + (1-\eta)E \text{ (kWh)}$$

where,

E is the total energy supplied in one year (kWh)
SG is the sundry gains from electrical devices, people and solar influx (kWh)
U is the overall U-value for the building, (W $m^{-2}K^{-1}$)

viz,

$$(1/UA)_{total} = (1/UA)_{walls} + (1/UA)_{glazing} + (1/UA)_{roof} + (1/UA)_{base}$$

n is the average number of ventilation air changes/hour
V is the enclosed volume of the building (m^3)
T_i is the annual mean inside air temperature (°C)
T_o is the annual mean outside air temperature (°C)
and η is the annual furnace/boiler efficiency.

A1.7 THERMAL COMFORT

The specifications for the inside and outside temperatures, T_i and T_o,depend upon considerations of the inside thermal comfort conditions required and the outside climate. A typical building in the UK will be maintained at around 20°C, whilst the mean annual outside temperature (at 51.7°N) is about 11°C. A reduction in mean annual inside temperature of 2°C (to 18°C) would result in a saving of two-ninths (22%) of the annual energy bill. Clearly, inside air temperatures should be specified accurately and controlled precisely.

A1.8 HEAT GAINS TO BUILDINGS

Sundry gains to buildings can sometimes comprise a high proportion of the input energy. In factories and kitchens, internal heat losses from processes often overwhelm the heat losses from the building fabric, resulting in the need for excessive fresh air ventilation to maintain the internal environment in a comfortable condition.

Solar gains

Solar gains to office blocks can also be high and, unless insolation is reduced by employing blinds, shutters or solar control glazing, such buildings may have to be air-conditioned.

Heat gains from people

In lecture theatres, auditoria, leisure centres and other public meeting places, the sensible and latent heat gains from people may dominate the heat balance equation. The sensible and latent heat proportions depend upon the rates of working, the internal environmental dry-bulb temperature and the relative humidity of the air. If not known, occupation densities could be assumed as follows:

10 m^2 in an office block
20 m^2 in executive offices
2 m^2 in restaurants
0.5 m^2 in cinemas and theatres

Heat gains from electrical devices

All electricity supplied to a building is a propensive sundry gain, which aids heating systems and combats cooling arrangements. The exceptions to this include electricity used for refrigeration, compressed air, extract fans and domestic hot water. Electricity for lighting often constitutes a major cooling load, especially in deep plan buildings. All the power supplied to electrical motors is eventually dissipated as heat. Heat dissipations from processes might be sensible and/or latent. In constructing an energy audit, sundry heat gains from people, solar gains, electricity utilisation and other fuel-using processes must be quantified. For a typical domestic dwelling, for example, the annual energy release for non-heating purposes can amount to an equivalent 1kW of continuous power rating.

Table A1.6 contains recommended airchange rates for some locations and activities.

A1.9 CARBON TAXES

The European Union has proposed carbon taxes as follows: £18/tonne of carbon emitted in 1993 to £60/tonne of carbon emitted in the year 2000.

Natural gas (mainly methane, CH_4, releasing 50 MJ/kg)

from combustion chemistry,
16 kg of natural gas releases 44 kg of CO_2
1 kg of natural gas releases
2.75 kg of CO_2 (0.75 kg of carbon) and 50 MJ (= 14 kWh)

Thus 1 kWh is provided by 0.072 kg of natural gas, whilst 0.198 kg of carbon dioxide, containing 0.054 kg of carbon, is released to the atmosphere in the process.

Table A1.6 Recommended airchange rates

Application	Litres/person	Litres s^{-1} m^{-2}	Occupancy unknown airchange/hr
Private dwellings	8–12	–	–
Board rooms	18–25	–	–
Bars	12–18	–	–
Stores	6–8	–	–
Factories	16–28	0.8	–
Garages	–	8.0	5
Operating theatres	–	16.0	–
Hospital wards	8–12	–	–
General offices	6–8	1.3–2.0	3–8
Private offices	8–12	1.3–2.0	3–8
Restaurants	12–18	–	5–10
Theatres, cinemas	6–8	–	5–10
Schools	14	–	–
Engine rooms	–	–	4
Baths	–	–	5–8
Lavatories	–	–	5–10
Kitchens	–	–	10–40

Gas price rises?

1993	**2000**
0.0972p/kWh	0.324p/kWh

Coal (mainly carbon, releasing 28 MJ/kg)

from combustion chemistry,
12 kg of carbon releases 44 kg of CO_2
1 kg of carbon releases 3.67 kg of CO_2 (1 kg of carbon) and 28 MJ (= 7.8 kWh)

Thus 1 kWh is provided by 0.128 kg of carbon, whilst 0.471 kg of carbon dioxide, containing 0.128 kg of carbon, is released to the atmosphere in the process.

Coal price rises?

1993	**2000**
0.231p/kWh	0.768p/kWh

Oil $\approx$ CH_2, releasing 45 MJ/kg)

$10/barrel of oil by the year 2000 (6300 MJ = 1750 kWh) = 1.14 p/kWh
from combustion chemistry,

14 kg of oil releases 44 kg of CO_2
1 kg of oil releases 3.14 kg of CO_2 (0.86 kg of carbon) and 45 MJ (= 12.5 kWh)

Thus 1 kWh is provided by 0.08 kg of oil, whilst 0.251 kg of carbon dioxide, containing 0.069 kg of carbon, is released to the atmosphere in the process.

Oil price rises?

1993	**2000**
0.123p/kWh	0.414p/kWh

Electricity

1 kWh of electricity requires approximately 3 kWh of coal, or 0.384 kg of carbon to be burnt at the power station. When combusted, this releases 1.4 kg of carbon dioxide, containing 0.384 kg of carbon, to the atmosphere.

It is not clear whether the power generator will be liable pay carbon taxes for this release, which, for each kWh of electricity, would amount to:

Electricity price rises?

1993	**2000**
0.69p/kWh	2.304p/kWh

As yet, these carbon taxes have not been enforced in the UK.

Bibliography

Allaby, M., *The Survival Handbook*, Pan, London, 1975.

Avallone, E.A. and Baumeister III, T., *Marks' Standard Handbook for Mechanical Engineers*, 9th Edition, McGraw-Hill, New York, 1987.

Blasch, I., Rasti, A. and Symvoulidou, M., *Integrated Environmental Management of Sudbrook Paper Mill*, Cranfield University, 1994.

Boyer, S., Grace, M. and Price, T., *International Combustion Ltd., Energy and Environmental Audit Report*, Cranfield University, 1994.

BP Statistical Review of World Energy, Corporate Communications Services, London, 1970–1991.

Breathing in Our Cities, Parliamentary Office of Science and Technology, House of Commons, 1994.

British Standards Institution, *BS5750, Specification for design, development, production, installation and servicing Quality Systems*, 1987.

Brown, A., *The U.K. Environment*, Department of the Environment, Government Statistical Service, HMSO, London, 1992.

Bruntland Report, World Commission on Environment and Development, *Our Common Future*, Oxford University Press, Oxford, 1987.

Button, J./Friends of the Earth, *How to be Green*, Century Hutchinson, 1989.

Chlorofluorocarbons—Professional and Practical Guidance, Guidance Note, CIBSE, London, December 1989.

Collin, P.H., *Dictionary of Ecology and the Environment*, Peter Collin Publishing, Teddington, 1988.

Department of Energy, *An Evaluation of Energy Related Greenhouse Gas Emissions and Measures to Ameliorate Them*, Energy Paper Number 58, U.K. Country Study for the Intergovernmental Study on Climate Change Response Strategies Working Group, Energy and Industry Sub Group, HMSO, London, October 1989.

Department of the Environment, *Digest of Environmental Protection and Water Statistics*, No. 15, HMSO, London, 1992.

Department of the Environment and the Welsh Office, *Integrated Pollution Control: A Practical Guide*, HMSO, London, 1993.

Department of the Environment, *U.K. Strategy for Sustainable Development*, Consultation Paper, HMSO, London, July 1993.

Doble, M., Energy and Environmental Audit of Keunen Brothers Tannery, MSc thesis, Cranfield University, 1992.

EC Council Regulation allowing voluntary participation by companies in the industrial sector in a Community eco-management and audit scheme, Council Regulation (EEC) No. 1836/93, *Official Journal of the European Communities*, No. L 168/1, 10.7.93.

Ehrlich P. and Ehrlich, A., *The Population Explosion*, Arrow, London, 1991.

Elfaki, M.E., Energy Audit of British Oxygen Company, MSc thesis, Cranfield University, 1993.

Elkington. J. and Hailes, J., *The Green Consumer Guide*, Gollancz, London, 1989.

Elkington, J. and Hailes, J., *The Green Consumer's Supermarket Shopping Guide*, Gollancz, London, 1989.

Embar, L.R., Layman, P.L., Lepkowski, W. and Zurer, P.S., Tending Global Commons—The Changing Atmosphere, *Chemical and Engineering News*, **64**, 47, 16–35, November 1986.

Energy 2000: A Global Strategy for Sustainable Development, World Commission on Environment and Development, Zed Books, 1987.

Environmental Protection (Applications, Appeals and Registers) Regulations, HMSO, London, 1991.

Environmental Protection (Authorisation of Processes) (Determination Periods) Order, HMSO, London, 1991.

Environmental Protection (Prescribed Processes and Substances) Regulations, HMSO, London, 1991.

Environmental Protection Act, HMSO, London, 1990.

Frengler, F., Water Management, MSc thesis, Cranfield University, 1993.

Government White Paper, *This Common Inheritance*, HMSO, London, 1990.

Gribbon, J., *Hothouse Earth—The Greenhouse Effect and Gaia*, Bantam Press, London, 1990.

Her Majesty's Inspectorate of Pollution, *Chief Inspector's Guidance to Inspectors, Environmental Protection Act 1990*, IPR Process Guidance Notes, HMSO, London, 1991.

HMSO, *Greenhouse Effect, Report of House of Lords Select Committee on Science and Technology, Volume II—Evidence*, HMSO, 1989.

Houghton, R.A. and Woodwell, G.M., Global Climatic Change, *Scientific American*, **260**, 4, 18–26, April 1989.

Kenny, P., Water Conservation and Auditing, MSc thesis, Cranfield Institute of Technology, 1992.

Kenny, P., Water Management, MSc thesis, Cranfield University, 1992.

Kirkpatrick, N., *Life Cycle Analysis and Eco-labelling*, PIRA International, Randalls Road, Leatherhead, Surrey, January, 1992.

Lovelock, J., *The Ages of Gaia*, Oxford, 1989.

March Consulting Group, *Reducing Costs and Improving Performance through Waste Minimisation—'The Aire and Calder Experience'*, Version 1.2, Manchester, 1994.

Marchewka, R., Rain-water Harvesting, MSc thesis, Cranfield Institute of Technology, 1993.

McElroy, M., The Challenge of Global Change, *New Scientist*, 34–36, 28 July 1988.

O'Callaghan, P.W., *Building for Energy Conservation*, Pergamon Press, Oxford, 1978.

O'Callaghan, P.W., *Energy for Industry*, Pergamon Press, 1978.

O'Callaghan, P.W., Efficient Use of Industrial Energy, paper presented at Seminar on *Economics of Industrial Energy*, University of Coimbra, Portugal, May, 1980.

O'Callaghan, P.W., Energy Audit, paper presented at *Energy Seminar*, AERE Harwell, September 1980.

O'Callaghan, P.W., Energy Management, *Retrofitting for Energy Conservation Handbook*, Ed. Milton Meckler, Marcel Dekker, New York, 1980.

O'Callaghan, P.W., *Design and Management for Energy Conservation*, Pergamon Press, Oxford, 1981.

O'Callaghan, P.W., *Energy Management*, McGraw-Hill, London, 1993.

O'Callaghan, P.W. Energy Resources, CO_2 Production and Energy Conservation, paper presented at Cranfield Triple E Seminar, *Energy, Environment, Ecology*, 25 September 1990.

Porteous, A., *Dictionary of Environmental Science and Technology*, Open University Press, Milton Keynes, 1991.
Reid, D., Tessier, L. and Thomas, P., *Integrated Environmental Management of IMI Refiners*, Cranfield University, 1994.
Rogers, G.F.C., Energy Conservation—choice or necessity?, *Chemical and Metallurgical Engineering*, Institution of Mechanical Engineers, 65–79, May, 1975.
Sadgrove, K., *The Green Manager's Handbook*, Gower, Aldershot, 1992.
Severn Trent Water, *Water Supply*, Severn Trent Information Services, January 1990.
Shone, M.O., Waste Reclamation and Waste Auditing, MSc Thesis, Cranfield Institute of Technology, 1992.
Shrewsbury, C., Telfer, J., Olsen, H. and Daskalopulos, N., *Energy and Environmental Audit: Boots the Chemist*, Cranfield University, 1994.
Smith, P., Royal Institute of British Architects Conference Discusses Greenhouse Effect, *Energy Management*, 18–21, UK. Department of Energy, April/May 1990.
Stikker, A., Sustainability and Business Management, *Business Strategy and the Environment*, **1**, 3, Autumn, 1992.
Taylor, A.E.B., Energy and Environmental Policies for Agriculture, MSc Thesis, Cranfield Institute of Technology, 1991.
The Bedfordshire Environment, Reports 1 and 2, Bedfordshire County Council, 1991, 1992.
U.K. Digest of Statistics, UK Statistical Office, London, 1990.
W.S. Atkins/March Consulting Group/Aspects International, *Project Catalyst, Report to the Demos Project Event*, Manchester Airport, 27 June 1994.
WARMER Factsheet, *Fuel From Waste*, The World Resource Foundation, Tonbridge, UK, January, 1991.
Water Act, HMSO, London, 1989.
Water Industry Act, HMSO, London, 1991.
Water Resources Act, HMSO, London, 1991.
Water Services Association, *Water Wise—Household Guide to Water*, Water Information, London, December 1991.
Welford, R. and Gouldson, A., *Environmental Management and Business Strategy*, Pitman Publishing, London, 1993.
Wright, G. and Wright, D., *Kingfisher Atlas*, Kingfisher Books, London, 1983.

Index

39 steps of environmental management 140
39 steps to optimise energy management investments at a site 172
39 steps to optimise waste management investments at a site 274
39 steps to optimise water management investments at a site 242

accidents 99
acid rain 1, 3–4, 11, 25
acidification 37
Acts and Regulations 38
aeration 235
aerosols 5
Agenda 21 22
agricultural effects 3, 9
air conditioning 191, 203
air pollution 2–4, 38, 90
air quality 37
airchange rates 356
algae blooms 9, 14, 224
Alkali, etc. Works Regulation 45
application of IPC authorisation 48
asthma 163
audit cycle 101
auditor 101
autoclaves 196

baths 250
BATNEEC, general 58
Best Available Techniques Not Entailing Excessive Cost (BATNEEC) 27, 33, 45–6, 58, 116
best investments 209
best practicable environmental option (BPEO) 33, 45
best practical means (BPMs) 27, 33
biodiversity 3–4, 24, 37
biological oxygen demand (BOD) 224
black list 227
black smoke 11
boiler efficiencies 352
boiler efficiency report 352
boilers 190, 196, 350
boilers, energy conservation in 351
BS 5750—quality systems 28, 80
BS 7750—environmental management systems 28–9, 80, 83, 87
building heat balance 353
building services 93

calorie 17
calorific values 148
calorific values of organic wastes 271
carbon cycle 7, 342
carbon dioxide 5, 9–10, 20, 155–6, 342
carbon monoxide 4, 11, 20, 157
carbon taxes 10, 355
catalytic convertors 11
chemical oxygen demand 224
Chief Inspectors Guidance Notes 47
chilled water 192, 247
chlorine 12
chlorofluorocarbons (CFCs) 5, 7–8, 12, 165
clarification 235
Clean Air Acts 38
clean technologies 42
climate change 11, 37
climate models 7
co-operation procedure 35
coagulation 235
coal 10, 144, 194
coliforms 226
combined heat and power 25
combustion 4, 143, 148
combustion calculations 151
compliance register 80

composting 267
compressed air services 192, 204
conservation 29
conservationism 23
consumerism 18, 255
Control of Pollution (Special Wastes) Regulations 43
Control of Pollution Act 39, 42
controls 209, 247
cycles of life 342

decay 10, 34
definitions 101
deforestation 3, 37
depletion of resources 3, 4
desertification 3, 9, 24
desulphurisation 11
diesel engines 11
dish washers 251
disinfection 236
driers 197
drought 3, 9, 24
duty of care 27, 34, 45

Earth Summit 22
EC Action Programmes 33, 153, 222, 258
EC Directives 23, 33–4, 153, 258
EC Environmental Action Programmes 24, 35
eco-audits 37
ecolabelling 2, 25, 37, 83
ecological effects 3
ecology 98
ecomanagement and audit scheme (EMAS) 28, 77, 80, 87
economics 2
ecosphere 1
effluent charges 29
effluents 2, 220
effluents to water 34
electrical services 191
electricity 194
emissions to air 34, 94, 143, 152
emissions to water 34
end of pipe clean up 24
energy 2, 143
  account 93, 171, 173
  and air pollution 25
  audit chart 183
  clean 3
  conservation 12, 189
  consumption 16, 188
  costing 21
  equations 349
  improvement options 182
  management 79, 90, 143, 168
  management checklists 194
  management procedures 188
  monitoring 189
  questions for the client interview 188
  reclamation 186
  recovery 26
  rules for the efficient conservation of 183
  saving options 215
  signature 305
  storage 190, 199
environment 101
environmental accounting 90, 101, 122
environmental audit 21, 79, 102, 104, 130
environmental commitment 109
environmental compliance audit 103, 116
environmental costing 21
environmental education 109
environmental effect 83, 103
environmental effects register, register of environmental effects 80, 103, 120
environmental engineering 2, 88, 90, 104
environmental impact 17, 23
environmental impact assessment 36, 104
environmental improvement options 129
environmental issues 90
environmental law 26, 34
environmental legislation 29, 33
environmental liability audits 106
environmental management 2, 88, 90, 104
environmental management manual 82, 104
environmental management programme 82, 104
environmental management review 104
environmental management system 28, 79, 105, 118
environmental objectives 92, 105, 117
environmental organisation 115
environmental policy 79, 90, 105, 112
environmental pollution 33
environmental probity of suppliers 82
environmental problems 3
environmental product audit 105
environmental programme 80
Environmental Protection (Applications, Appeals and Registers) Regulations 49

Environmental Protection (Prescribed Processes and Substances) Regulations 27, 46
Environmental Protection Act (EPA) 27, 34, 42–3, 45, 303
environmental review 79, 111, 120
environmental standards 2, 28–9, 77
environmental statement 80, 105, 112
environmental statutory nuisance 34
environmental targets 82, 105, 117
essential reference material 106
eutrophication 1, 14
exergy 186
exhaust emissions 20
extreme weather events 9

fabric transmission 215
fermentation 10
fertilisers 14, 17
filtration 235
financial practices 188
flocculation 235
floods 9, 11
flue gas desulphurisation equipment 11
food 262
food shortage 9
fossil fuel consumption 17
fossil fuels 3, 143
furnaces 189, 195

gas 144, 194
glacial melting 9
global environmental burden 23
global environmental problems 3
global temperature rise 6, 8
global warming 1, 3, 10, 18, 25, 29
  combating 10
  effects of 9
global warming potential 8
greenhouse effect 3, 14
greenhouse gases 5, 8
grey list 39, 227
grey water recycling 251
grit removal 237
gypsum 11

Haber process 14
halocarbons 6
halons 12
hardness 226
hazardous waste 43
health 99
heat gains to buildings 354
heating systems 198
Her Majesty's Inspectorate of Pollution (HMIP) 27, 33, 42, 45
HMIP authorisation 45
hot water systems 191, 204
hurricanes 9
hydrocarbons 4

ice age 8
impact 2
incineration 26, 43, 267
incomplete combustion 4
Industry Sector Guidance Notes 49
inert waste 43
information 101
insecticides 1
integrated case study 284
integrated environmental management 2, 87
integrated pollution control (IPC) 2, 25, 27, 33, 37, 45, 116
Integrated pollution prevention and control (IPPC) 33, 36, 43
investment opportunities 217
investment plan 217
investment portfolio 253, 281, 330
IPR Notes 49

land pollution 2–4, 90
lavatories 250
laws of energy and materials flows 280
leachate 4
legislation 2, 26, 29, 153, 258
liabilities audits 106
life cycle analysis 83
life cycle assessment 83
lighting 191, 200
lignite 10
Local Authority Air Pollution Control (LAAPC) 27, 33
local authority waste disposal companies (LAWDC) 33, 47
local environmental problems 3

materials 2, 97, 255
  options for improvement 266
  rules for the efficient conservation of 276
materials accounting 266
materials management 79, 90
materials reclamation 208

materials recycling 26
materials uses 266
methane 5, 14, 163
monergy audit chart 184
monitoring and targeting 253, 281
Montreal protocol 13
motor vehicles 20
municipal waste 43

National Rivers Authority (NRA) 27, 33, 41
nature 98
nitrate control 26
nitrates 3
nitric acids 4
nitrogen cycle 13, 345, 347
nitrogen oxides 4–5, 10, 20
nitrous oxide 158
noise 2, 37, 79
NOx emissions 25
nuclear explosions 4
nuclear waste 4
nuisance 34, 45, 98
  private 34
  public 34
  statutory 27, 34

oil 144, 194
  crude 10
  fuel 10
oil pollution 3
organic materials 9
organic waste 14, 150
organic waste processing 271
orimulsion 10
over-farming 3
ozone 5, 10
  low-level 11, 26
ozone layer 18
ozone layer depletion 2–4, 8, 12, 24, 29, 37
ozone-depleting chemicals 12

packaging 43, 79, 100
particulates 11
penalties for pollution offences 28, 48
personnel energy awareness 189
petrol 10
pH adjustment 236
pH level 226
phosphorous cycle 14, 345, 348
photochemical smog 5
photosynthesis 7, 8, 10, 342
phytoplankton 1–2, 7, 12, 14
policy 2
pollution 25–6, 33, 164, 187, 255
pollution abatement technologies 36
pollution accounting 179
pollution taxes 29
population 3, 15, 25
power generators 10
precautionary principle 9
precipitation 9
prescribed processes 27, 33, 64, 154
prescribed substances 27, 33, 39, 74, 222
processes 91
product life cycles 99
product planning 79
products 2, 22, 99, 255
project plan 253, 281, 330
pyrolysis 267

radiation involving gases 350
rainforest destruction 3, 24
rainwater harvesting 251
reclamation 79
red list 39, 42, 231
refrigeration plant 192, 207, 247
refuse-derived fuels 272
regulations 25–6, 153
renewable economy 23
resource conservation 29
respiration 7, 10, 344
respiratory problems 4, 10
reuse 22, 42, 79, 267–8
river quality classifications 230
road transport 159

safety 99
scoping study 108, 117
screening 235–6
sea level rise 9
sedimentary cycles 345
sedimentation 237
self-sufficiency 18
services 92
sewage 14, 26
showers 250
Single European Act 35
smog 2, 5, 11
smoke 4, 11
solar conversion efficiencies for various crops 346
soot 4, 11

space heating 190, 198
steam plant 192, 207
storm surges 9
subsidiarity 24, 35
substances 92
sulphur 10
sulphur dioxide 4, 10–11
sulphur oxides 160
sulphuric acid 4, 8, 10
surfactants 226
sustainable development 22, 36

tap flow restrictors 249
taps 249
tariffs 194
thermal comfort 354
thermal insulation 200
throwaway society 21, 256
toilets 250
Toronto protocol 10
toxic substances 39
traffic pollution 4
transportation 18, 93
tropical diseases 9
turbidity 226

U-Value 349
UK policy 24
unburnt hydrocarbons 4, 11

validation of the environmental management system 138
vegetation 14
vehicle pollution 25
ventilation 191, 201
volatile organic compounds 20, 25, 162
volcanic eruptions 6, 8, 10, 12

washing machines 251
waste 2–4, 13, 26, 37, 43, 45, 97
  accounting 266
  and the law 257
  arisings 21, 266
  avoidance 79
  calorific values of 271–2
  categories 261
  civic amenity 43
  collection authorities (WCA) 34, 42, 47
  combustible 262
  consignment systems 43
  controlled 43
  definitions 43
  difficult 43, 263
  disposal 25
    authorities (WDA) 27, 34
    charges 29
    options 267
  elimination 267–8
  energy from 270
  food 262
  heat recovery 208
  improvement project 281
  incineration 26
  inert 261
  management 90, 255
  management checklist 278
  materials, questions for the client interview 276
  minimisation 42
  minimisation options 267
  on land 34, 42, 47
  processing 271
  prohibited 265
  putrescible 43, 262
  recycling 22, 26, 42, 79, 267, 268
  regulation authorities (WRA) 27, 34, 41, 47
  semi-inert 43, 262
  special 43, 264
  water 236
wastes, notifiable 43
wastes, prohibited 43
water 37, 220
  checklists for water managers 246
  discharges to 96
  domestic hot water 247
  drinking 234
  Drinking Water Inspectorate 42
  drinking water quality standards 228
  eutrophication of 1, 14
  questions for the client interview 244
  rules for the efficienct conservation of 243
  uses for 239
water accounting 95, 241
Water Act 27, 38, 41
water and the law 222
water authority 39
water companies 41
water conservation options 248, 252
water cycle 13, 220

water improvement project 253
water industry 237
Water Industry Act 28, 42
water leakage 248
water management 42, 79, 90, 220, 239
water metering 248
water pollution 2, 4, 25–6, 38, 41, 90, 221
water prices 238
water quality 3, 224
water quality standards 26
water recycling 251
Water Resources Act 28, 42
water shortage 9
water softening 236
water supplies 220
water treatment 234
weather events 9